Skycrane

Igor Sikorsky's Last Vision

Skycrane

Igor Sikorsky's Last Vision

John A. McKenna
Retired Executive Vice President
Sikorsky Aircraft

LIBRARY OF FLIGHT

Ned Allen, Editor-in-Chief
Lockheed Martin Corporation
Bethesda, Maryland

Published by
American Institute of Aeronautics and Astronautics, Inc.
1801 Alexander Bell Drive, Reston, VA 20191-4344

Library of Congress Cataloging-in-Publication Data

McKenna, John A.
Skycrane : Igor Sikorsky's last vision / John A. McKenna ; Ned Allen, editor-in-chief.
p. cm.
Includes index.
ISBN 978-1-60086-756-9
1. Sikorsky helicopters--History. 2. Sikorsky, Igor Ivan, 1889–1972. I. Allen, Ned. II. Title.
TL716.9.S55M34 2010
629.133'352--dc22 2010019965

Cover design by Virginia Kozlowski; front cover photograph © Alfredo La Marca.

DEDICATION

To the following three extraordinary men:

Igor Sikorsky,
who invented the Skycrane,

General Harry Kinnard,
who chose the Skycrane to support combat operations in Vietnam, and

Jack Erickson,
who developed commercial use of the Skycrane

A Note from the Editor-in-Chief

This case study of the Skycrane was selected for publication within the *Library of Flight* for three reasons: it paints a vivid picture of the last creative effort of one of aerospace's great geniuses; it shows that aerospace genius is a profoundly entrepreneurial trait, not just a technical or academic one; and it is well-crafted and engagingly written by John A. McKenna.

Today many large and maturing aerospace businesses, facing building pressure to provide more affordable products, are concerned with restoring to their operations the verve and vigor they imagine our industry's pioneers possessed. Many look to re-sparking invention (coming up with new ideas) and innovation (transitioning ideas into products) within their organizations as a formula for rejuvenation. Many lament the added complexity of our large businesses today—complexity that appears to block innovation with a no man's land of bureaucratic obstacles and required approvals separating the innovator from her or his vision. The Skycrane was conceived when Sikorsky Aircraft was already a large and mature business, housed within the even larger United Technologies Corporation (then United Aircraft Corporation), yet the Skycrane story shows that gifted inventor–entrepreneurs can and do succeed. As Mr Sikorsky said: "… the individual still remains the spark that moves mankind …"

Ned Allen
May 2010
Bethesda Maryland

FOREWORD

The name of Igor Sikorsky is most frequently associated with the birth and development of the worldwide helicopter industry. Often overlooked is the fact that my father pioneered three phases of aviation:

- He designed and flew the world's first four-engine airliner at the age of 24. This remarkable aircraft first flew in May 1913. In August it established a world record by completing a flight of one hour and fifty minutes, while carrying eight passengers in its roomy cabin. It is indeed remarkable, when one considers, that this large aircraft flew a short nine and a half years after the Wright brothers' first flight at Kitty Hawk.
- During the Bolshevik revolution, Igor Sikorsky left Russia and settled in the Unites States. In the early 1930s he designed a series of flying boats with which Pan American Airways created airline routes across the Caribbean deep into South America and pioneered trans-Pacific and trans-Atlantic airline service.
- Igor Sikorsky's third career was marked by the successful development of the single-rotor helicopter. The prototype VS-300 of 1939 to 1941 led to a number of larger, more powerful helicopters in the following years.

The S-64 Flying Crane was the last helicopter design that my father initiated. The concept was greeted with some skepticism when he first proposed it in the 1950s. I well remember conversations with my father that led to the quick sketches I drew in 1957, showing some of the applications we envisioned for a Crane demonstrator. Slowly but steadily, he convinced his colleagues and engineers. The prototype S-60, proof of concept Crane, flew in March of 1959 and led to the S-64, which made its first flight in May of 1962.

The author of this book is well-qualified to record the birth and development of the Crane concept. John "Jack" McKenna was responsible for building the first S-64 in 1961 and for supporting the follow-on Army and commercial programs as the executive vice president at Sikorsky Aircraft. Skycranes are performing as commercial heavy lifters and firefighters in America, Europe, and Asia today. I am convinced that many more missions for the Crane are waiting to be discovered.

Sergei I. Sikorsky

Contents

Preface

Several years ago, Ray Leoni, the author of *Black Hawk: The Story of a World Class Helicopter* published by AIAA, asked me to review a part of his book because I had worked on development of the Black Hawk early on. At the end of that conversation, I congratulated him on his book and commented that I had always wanted to write a book about the Skycrane. With his enthusiastic encouragement, I proceeded.

There were several motivations. The Skycrane was the last creation of Igor Sikorsky, who was one of the great innovators of the aviation world. The full Skycrane story, however, had never been told. It was a radical design that many thought would not be successful. It was also the last creative effort of a man considered to be an aircraft design genius and the greatest gentleman that I have ever known.

I was responsible for building his first S-64 Skycrane, and I recall standing on the factory floor under this barely assembled, immense helicopter. As I looked up to the bottom of the fuselage, which was nine feet off the factory floor, and at the aft-facing pilot's cockpit, I worried that Mr. Sikorsky's unique concepts for a cargo helicopter might not succeed with this strange vehicle.

There was no written Army requirement for the Skycrane, but once it demonstrated its skills to rapidly and precisely move critical Army cargo, the wisdom of Mr. Sikorsky's design was revealed. The Army placed an order for six Skycranes to perform a detailed Army evaluation. This evaluation was completed in two months, and Skycranes were sent immediately to Vietnam. They proved to be valuable, rugged, and reliable support for high-priority combat operations. A production order followed, so more Skycranes could be deployed to Vietnam. As the Army retired its Skycranes, about 30 years later, many were converted to successful commercial use.

This book is about the many people who helped Mr. Sikorsky's Skycrane become so successful. Skycranes are performing valuable and urgently needed commercial services on a daily basis in America, Europe, and Asia, 48 years after the first flight of the Skycrane. It is hard to identify any other aircraft that first flew 48 years ago that is currently saving lives, providing essential services, and earning profits on a daily basis, worldwide. These factors make for a great story.

The book is dedicated to the three men who led the Skycrane through the three phases of its development. Initially, **Igor Sikorsky** got sparse support inside and outside the company for the Crane concept. He continued with speeches in 1955 to potential military and commercial users in the United States and the United Kingdom, even a meeting with the logging industry. He was proved correct, as shown in Chapter 7, which discusses the first commercial Skycrane (purchased for logging).

After a campaign of several years, Mr. Sikorsky obtained company funds to build an experimental Crane, using obsolete hardware. Its flight tests in 1959 proved the concept, but produced no orders. As he continued to develop the Crane concept, interest grew, and he obtained company funding to build and flight test a more expensive, all-new Crane in 1961. It was the largest helicopter in the free world, with no promised orders.

A few years later, General **Harry Kinnard** decided the Army needed the powerful lift of the Skycranes. He was a highly decorated combat paratrooper, who jumped into France on D-Day and commanded troops in the Battle of the Bulge in Germany. As commander of the new Air Assault Division in 1962, he was developing the groundbreaking concept of moving troops into combat and supporting them with helicopters instead of trucks.

He was given the Army-standard Huey and Chinook helicopters, but insisted on adding the higher payload Skycranes to ensure the troops had the necessary artillery, ammunition, fuel, food, and vehicles to succeed. The Army ordered Skycranes in 1963, and they flew to Fort Benning in 1964 to train with his new division. Two months later he took them with his division to Vietnam to support combat operations with great success. Army production orders followed.

The first commercial Skycrane was bought in 1971, ten years after first flight, by **Jack Erickson** of Erickson Logging Company. This Skycrane moved high-value timber from the remote forests of Oregon, where Jack learned his logging skills. He made many valuable changes in the Skycrane, and it became profitable. He also developed applications for firefighting and for building electric power lines in remote areas.

When Army orders ended, he bought the world manufacturing and support rights for the Skycrane. Later, as Army Skycranes were retired, he purchased many to convert them for his commercial use. He expanded his fleet to almost twenty Skycranes and also sold rebuilt Army Skycranes to other commercial helicopter users. His company continues to improve the original Skycrane design and support it worldwide.

John A. McKenna
September 2010

Acknowledgments

This book would not exist without the help of Daniel Libertino, president of the Igor I. Sikorsky Historical Archives, Inc. He did personal research in the Archives for many parts of this book and encouraged many other retired Sikorsky employees to assist me repeatedly. All of these people were extremely helpful, even though their real task is to organize thousands of articles, reports, photographs, and other documents into a tremendous historical library of Igor I. Sikorsky and Sikorsky Aircraft.

The Archives are located in the original farmhouse on the Sikorsky Aircraft property in Strafford, Connecticut (www.sikorskyarchives.com). They also have created and maintained an extensive museum next door to the Archives. Visitors are welcome, but must call the Archives in advance at 203-386-4356. Visitors will be able to see, firsthand, hundreds of models, photographs, reports, and other Sikorsky memorabilia at the museum. I was struck by the 1930 picture of the Sikorsky Plant at Roosevelt Field, New York, to which I rode five miles, on my bike, to buy model aircraft kits when I was a boy.

Sergei I. Sikorsky, son of Igor I. Sikorsky, wrote the book *The Sikorsky Legacy*, which provided valuable information for Chapter 1 regarding his father's thoughts in building his first unsuccessful helicopters in 1909 and 1910, as well as his wide range of fixed-wing aircraft in Russia and in America. His book explains the difficulties in finally developing the first successful helicopter in 1939 and the success with later helicopters. His sketches also help to explain Chapter 1.

Sergei also assisted in Chapter 4 regarding the West German Skycrane test program, by drawing from his experience in a series of marketing and management roles at Sikorsky Aircraft and in Europe. Ulrich Heider, the project manager for that two-year program, provided essential narrative and insight.

At my request, Sergei introduced me to Jack Erickson, president of the Erickson Air-Crane Company, who asked Clive Whittenbury to write about Erickson. Clive has a Ph.D. in aeronautical engineering with experience in several companies before joining Erickson, where he is now on the Board of Directors. He wrote an excellent Chapter 7, which I edited and was then approved by Jack Erickson. Clive also answered my many questions regarding Erickson's operations. Lee Ramage, who flew Skycranes at Sikorsky and Erickson before retiring as chief operations officer at Erickson, provided much valuable information regarding commercial Skycrane operations for Chapter 7.

Special thanks go to Lewis Knapp, retired head of Sikorsky Advanced Design, who recalled his daily work with Mr. Sikorsky as they worked together to complete the preliminary design of the S-60 (see Chapter 2) as well as his discussions with the experimental pilots on the S-60 flight, which carried people on the open wooden platform. Thanks also goes to Knapp for his recollection of how Mr. Sikorsky came to him again, to begin design of the S-64 and why George Howard decided to eliminate all of the fairings and work platforms on the upper deck of the S-64 fuselage (see Chapter 3).

Mal Burgess, the operations manager of the Skycrane commercial sales team, made Chapter 6 come alive as he decribed how the team invented solutions for commercial Skycrane demonstrations for potential customers in North and South America. He also provided valuable insight into the Sikorsky and Erickson organizations and operations, when later in his career he was vice president of operations at Erickson Air-Crane.

Thanks also go to Art Kesten, retired founder and publisher of *Army Aviation* magazine, who introduced me to the current publisher Bill Harris, who offered to run an ad in his magazine requesting information for Chapter 5 from Army pilots who flew the Skycrane in Vietnam. Sixteen Army pilots and crew members sent me over 300 pages of photos and documents on the CH-54 engineering development and Vietnam accomplishments, which were a tremendous contribution to a thorough Chapter 5.

Jay C. Rickmeyer, the lead Army engineer on the development phase of the Army CH-54, provided valuable documents, photographs, and narrative for Chapter 5, regarding the rapid development, production, and deployment to Vietnam of the first YCH-54As.

Lieutenant Colonel (retired) Gary R. Heffner, commander of the first company of production CH-54As in Vietnam, provided valuable data from the extensive records he kept of Skycrane operations. He also shared many reports of operational problems, photographs of CH-54 Vietnam flight operations, and his thoughts on the need for a more capable Crane to replace the Skycrane. His reviews of Chapter 5 also added accuracy and focus.

Lieutenant Colonel (retired) Warren R. Silva provided an excellent article on the recovery of a downed CH-54 and personal stories of several unique applications of the Skycrane. Lieutenant Colonel (retired) Eldridge W. Brock, commander of a CH-54A Aviation Company, reported his forced landing in a combat area, a Skycrane accident resulting in a lost aircraft, and his thoughts on a more capable Crane.

Chief Warrant Officer (retired) James R. Oden provided insight regarding the flight training of the first Army pilots in the CH- 54 and his unique experiences in Vietnam, including the dropping of 10,000-pound bombs from his CH-54. Lieutenant Colonel (retired) Paul J. Fardink, a CH-54 pilot and a writer of many aviation articles, provided his writings regarding the Skycrane world records and the difficulty the Skycrane had lifting the largest Marine Corps helicopter, which was downed in Vietnam. Raymond L. Robb, a contributing editor to *Vertiflite* magazine, provided several excellent photographs of a CH-54 on exhibit, which were used in the book.

Special thanks also go to Lee Jacobson, the retired Sikorsky director of aircraft design and development, who added valuable technical information, explained his helicopter gross weight theory and digitized over 100 photographs that I selected from the Archives. Thanks to Bill Tuttle, who proofread the entire original book and corrected my bad typing, syntax, and unclear text. Thanks also go to Art Linden, retired Sikorsky vice president of the Comanche Program, who provided valuable data and guidance for the three possible future Cranes as defined by material available in the Archives.

Finally, thanks go to Ray Leoni, retired senior vice president of engineering and advanced programs at Sikorsky Aircraft, who encouraged me to write the book when we were discussing the writing of his book, which became the successful *Black Hawk: The Story of a World Class Helicopter* published by AIAA. He also answered my many writing and publishing questions over two years. Thanks to Harry Hleva, retired Sikorsky product support manager, for providing me with a stirring *Readers Digest* article about Igor Sikorsky's determination to invent a successful helicopter.

Thanks go to my son John, who solved my many, many computer problems and added valuable ideas for the book. Thanks also go to my son Peter, who proofread final versions of the book and gave fresh focus in several areas. Thanks go to my wife Adele, who supported me through this long process.

John A. McKenna
September 2010

SKYCRANE TIMELINE

100 YEARS OF AVIATION HISTORY

1910	Igor Sikorsky gets a helicopter airborne by remote control.
1911	Igor “postpones” helicopter development.
1911	Igor designs and flies his airplane, the S-2.
1913	Igor designs and flies the world’s first four-engine aircraft.
1919	Igor flees the Russian revolution to go to America.
1920	Igor builds first U.S. aircraft outdoors.
1940	Igor builds the VS-44 four-engine flying boats.
1941	Igor flies world’s first successful single main rotor helicopter.
1942	Army accepts its first operational helicopter, a Sikorsky R-4.
1943	United States and United Kingdom order over 100 R-4s.
1945–1957	Igor takes his company to world leadership.
1955	Igor identifies new Crane helicopter concept.
1957	Igor retires as engineering chief.
1958	Igor leads S-60 experimental Crane design.
1959	First flight of S-60 experimental Crane.
1959	S-60 begins demonstrations at Army bases.
1959	S-60 proves Igor’s Crane concept is viable.
1961	Igor leads all-new S-64 production Crane design.
1962	First flight of S-64 production Crane.
1962	S-64 flies to Army bases to demonstrate abilities.
1963	West German military buys two S-64s to evaluate.
1963	Army buys six YCH-54As Cranes to evaluate.
1964	First delivery of YCH-54A is twelve months after order.
1964	Four YCH-54s sent to Fort Benning trials four months later.
1964	Four YCH-54As shipped to Vietnam two months later in Dec. 1964.
1965	Four YCH-54As begin combat support missions, moving artillery, ammo, fuel, trucks, bridges, medical pods, etc.
1965	FAA certification of CH-54A design, no longer a “Y.”
1966	CH-54A sets time to climb world’s helicopter record.
1967	S-64 offloads commercial container ship in Connecticut.

1967	Sales team takes S-64 to commercial customers.
1968	First six production CH-54As arrive in Vietnam.
1968	CH-54 offloads 224 tons from Aussie carrier in Vietnam.
1968	CH-54 offloads ship underway in Vietnam.
1968	443 CH-54 flights over design gross weight in Vietnam.
1968	Rowan Drilling leases two S-64s for Alaska oil work.
1969	CH-54Bs delivered with 25% higher payload than CH-54As.
1971	Erickson Logging buys first S-64 Skycrane.
1974	Other commercial operators buy Skycranes.
1995	Army retires all CH-54As and CH-54Bs.
1997	Erickson Air-Crane buys all rights to S-64 Skycrane.
1998	Erickson rebuilds its first Army surplus CH-54A.
2007	Erickson adds an eighteenth Crane to its fleet.
2008	Erickson flies one S-64 300 hours in 30 days of logging.
2008	Erickson sells its ninth S-64 to another user.
2009	Over 30 Skycranes flying on five continents.
2009	Erickson-refurbished CH-54 Skycrane sells for $28,000,000 vs $2,800,000 original cost to Army.
2010	Design of CH-53K is basis for possible 25-ton future Crane.

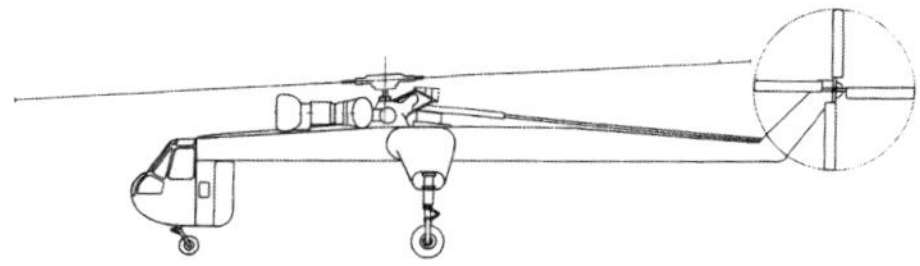

Chapter 1

Igor Sikorsky Defines the Crane Helicopter

The Russian Years

Igor I. Sikorsky was one of the greatest aviation innovators of the 20th century. His life in aviation began in Russia as described by his son, Sergei I. Sikorsky, in his book *The Sikorsky Legacy*.* Sergei writes that his father built his first helicopter, the H-1, in 1909 and another, the H-2, in 1910 (Fig. 1.1). The H-1 could not fly, and the H-2 could barely lift itself off the ground, but not with a pilot. His father said he would "temporarily postpone" further helicopter work, and so he moved onto the designing of airplanes.

Igor Sikorsky designed and flew his first airplane, the S-2, in 1911. In 1913, he designed and flew his airliner, called the Le Grand, the first four-engine aircraft to fly successfully. In 1913, he built the S-22 (Fig. 1.2), which was the largest four-engine seaplane in the world. The same year, he designed and flew an early fighter, the S-20.

Immigration and Success in America

After the Russian Revolution, Sergei said that his father came to America in 1919, arriving "in New York City with little money, no friends or business contacts and almost no English language capability," as shown in Fig. 1.3. In 1923, he began building his first American airplane outdoors, on the farm of a friend. In 1928, Igor Sikorsky then built his twin-engine amphibian, the S-38, and by 1942, the four-engine flying boats, including the VS-44 (Fig. 1.4), built for American Export Airlines with a 4000-mile range capability. This summary provides a brief glimpse of his tremendous innovations and success in single-engine and multi-engine land planes, amphibians, and flying boats.

Second Helicopter Attempt after 30 Years

During the 1930s, orders for Sikorsky flying boats dropped as the boats were replaced by lighter and lower drag land-based aircraft. Sikorsky planned

---

*Much of the information in this chapter comes from the book *The Sikorsky Legacy* © Sergei I. Sikorsky with the Igor I. Sikorsky Historical Archives, 2007, published by Arcadia Publishing of Charleston, South Carolina.

Fig. 1.1 H-2: Sikorsky's second helicopter lifted only a few inches in 1910, so development was "temporarily postponed."

to use the remaining resources of the Sikorsky organization not to close down the Sikorsky plant as instructed, but to invent the helicopter.†

This was Sikorsky's new goal, even though many of the noted pioneers of aviation had not succeeded with the helicopter, including himself in his original attempts of 1908 and 1909. While his company was building flying boats, Sikorsky started building and testing a helicopter again. In

†Harry Hleva, a retired Sikorsky product support manager, provided an article from the December 1956 *Readers Digest* entitled "The Most Unforgettable Character I've Met" written by Eugene R. Wilson, the executive to whom Igor Sikorsky and his company reported.

Fig. 1.2 S-22 seaplane.

September 1939, he hovered his first successful helicopter, the VS-300, as shown in Fig. 1.5. When asked why the early VS-300 could not fly forward, Sergei quotes his father as saying, "This is a minor technical problem that we have not solved yet."

Sergei related "that from September 1939 to December 1941, Igor Sikorsky tested and modified the VS-300, literally making hundreds of changes to the VS-300 before it also flew forward." He had developed this first practical helicopter after two years of flight testing to achieve forward flight. This opened the world to vertical flight! It also began a new chapter in world aviation and was the beginning of a huge new worldwide industry.

Fig. 1.3 Arrival in New York with little money or English capability in 1919.

ARMY ORDERS

The U.S. Army witnessed VS-300 flights and placed an order for an improved version, the XR-4 (shown in Fig. 1.6). In 1942, a single XR-4 two-place training helicopter was delivered to the Army. The Army quickly realized it had new capabilities. The U.S. Army Air Forces and British Forces then ordered over 200 of the production version, the R-4, and took delivery of 131, which were used in the Second World War (WWII). One R-4, for example, rescued three wounded British

Fig. 1.4 VS-44 four-engine flying boat.

commandos and an American pilot inside Japanese-held Burma. The delivery of the first XR-4 is shown in Fig. 1.7.

Sikorsky Aircraft Becomes the Free World's Largest Supplier

By the 1950s, Sikorsky Aircraft was the leading supplier of helicopters in the free world, with a new plant, which would grow to over 2,000,000 square feet. This was possible because of Sikorsky's genius and persistence. Well-qualified engineers and others were attracted to him and his company. He was involved in developing all of the helicopters designed and produced until his retirement in 1957. His seaplane experience can be seen in the boat hull design of the S-61, which first flew in 1959. With minor modification, an S-61 flies as the presidential helicopter today, 51 years after its first flight. As several new helicopters began production in the new plant, Sikorsky developed a vision for an entirely new type of helicopter with radical features, never seen before in any helicopter.

Crane Helicopter

In 1955, his earliest speech on record regarding the Crane, Sikorsky addressed a British engineering association, calling his vision the Cargo or Crane type as opposed to the Passenger type, saying, "It would be specifically designed to lift any object or load, including large and bulky ones, and to carry them on the outside, suspended below the body of the aircraft." He added, "The main difference would be in the body, location of the pilot's cabin, landing gear, and powerful, special lifting crane devices that would allow for picking up an object, if necessary, with the helicopter hovering over the ground."

Sikorsky also described the most radical feature of this helicopter, the role of the aft-facing pilot in his own rear-facing cockpit with special

a)

b)

c)

Fig. 1.5 VS-300: a) Michael Buivid (left) and Igor Sikorsky with component test rig circa 1932, b) Sikorsky hovers in early 1939, and c) flying forward to place the rod in the ring in 1941.

Fig. 1.6 VS-300 (foreground) flying with the XR-4.

flight controls as follows:

> this pilot would have excellent vision of the object which he is to lift, of the whole procedure of attaching and lifting and finally would have excellent opportunity to direct the aircraft in a way that would permit bringing and lowering the object...actually and precisely where it may be needed.

Fig. 1.7 Mr. Sikorsky (left) accompanied by Orville Wright (middle) delivers the XR-4 to the Army; it was accepted by Captain Gregory (right), who flew a B-17 to Wright Field, Ohio, to escort the XR-4.

Fig. 1.8 Sketch of a proposed experimental Crane.

The aft-facing pilot, shown in Fig. 1.8,[‡] has excellent visibility of the load and control of the helicopter to precisely place the radar control shed on this mountain location.

CREATING NEW INDUSTRIES

Sikorsky concludes his dissertation in London on the Cargo or Crane type by stating the following:

> It is this type of helicopter which will promote in the next decade or two a whole new series of enterprises and industries that would depend on

[‡]The drawings in this chapter were made by Sergei I. Sikorsky, who explains in the book *The Sikorsky Legacy* that "they were used by Igor Sikorsky to 'sell' the concept of a Crane using S-56 components. The helicopter in the sketches would become the S-60, which first flew in 1959."

> the possibility of making direct deliveries of objects, which owing to their length or bulk, could not otherwise be transported in assembled, ready to use form.

As an example of this new capability, Fig. 1.9 shows the Crane type delivering a power line tower.

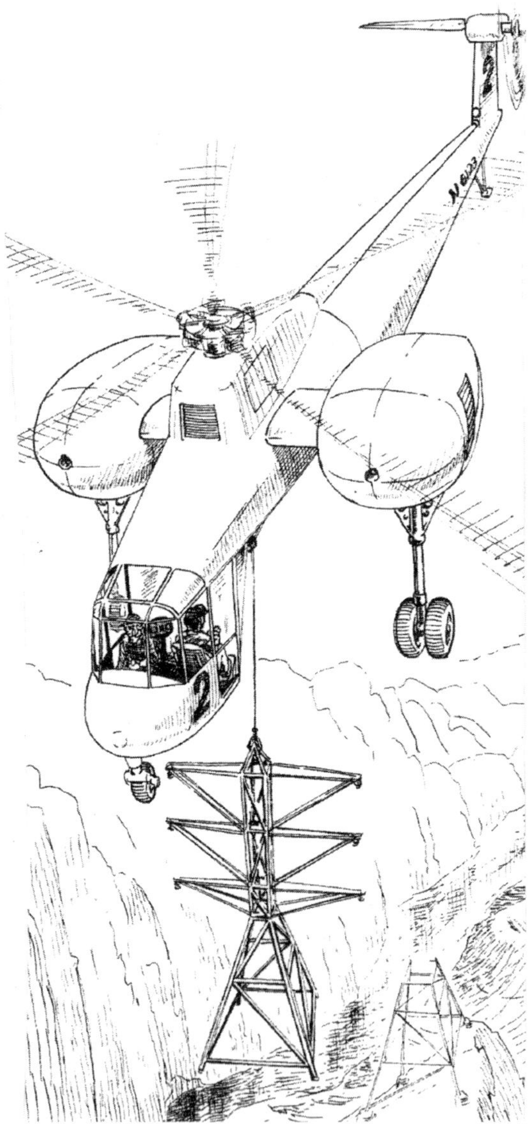

Fig. 1.9 Sketch of experimental Crane installing tower for electric power line in remote and difficult terrain.

Design Innovations

In this talk, he specifically listed the major design features of the Crane helicopter, dated its near-term arrival, and forecast wide benefits to new users and enterprises. At the time of this speech, Mr. Sikorsky was 68 years old, retired as chief of engineering of the company he founded, but still a consultant and active on a daily basis.

In the January 1958 issue of United Aircraft Corporation's *BEE-HIVE* magazine (p. 7), Frank Delear's article, "Tomorrow's Flying Cranes," quotes Mr. Sikorsky:

> The Crane helicopter has one unique and essentially important characteristic that places it far above the abilities of any other vehicle. There is no limitation as to the place where it may pick up cargo or the place where it may lower it. It has no limit, either as to the size or bulk of the object to be carried providing the weight does not exceed the aircrafts capabilities, thus it does not matter what the bulk of the cargo is or how inaccessible the spot of departure or arrival.

Delear reports that Mr. Sikorsky listed three vital advantages of the Crane helicopter in order of importance as follows (pp. 7–8):

1) Completely unobstructed visibility for the pilot, allowing him to observe the entire operation of picking up or delivering cargoes. This would mean windows all around and probably dual pilot seats—one facing aft and one facing forward.
2) A configuration for picking up while hovering, but also attaching such objects as trucks and tanks, or very long objects such as posts, lumber, and piping to the aircraft while it is parked on the ground.
3) A greater payload and, in equivalent numbers, a lower production cost than the present cabin helicopter.

There were many skeptics of the feasibility of the Crane helicopter because it was based on several new and radical features, including the following:

- An aft-facing pilot who would control the flight of the helicopter at the most critical point of picking up or delivering cargo of widely different sizes, shapes, and weights, and monitoring it during highly variable fight conditions
- A landing gear high and wide enough to taxi over a standard commercial shipping container or other large loads
- An adjustable landing gear, which would allow the Crane to squat over other loads to make direct hookup to the fuselage using several small special winches

- A main hoist capable of rapidly lifting multiton loads from 100 feet below the Crane
- No passenger cabin, but a *people pod*, readily attachable, that would permit transport of people and/or sensitive equipment in a cabin at much lower vibration levels than any other helicopter

Further Speeches Announce S-60 Crane

Sikorsky not only spoke at technical meetings, but also to industry groups. In a presentation to the National Meeting of the Forest Products Research Society in 1958, Mr. Sikorsky said the following[§]:

> The lumber industry is a vast and important one. It is believed that the helicopter could offer it a completely new line of important services... a special Crane helicopter is being designed and should be ready for industry use in several years. It should be able to fly a 6-ton load externally at 85 mph.

Because he was talking to people who managed businesses, he estimated the cost of a typical mission in 1958 dollars could be 93 cents per ton-mile and added, "As soon as jet engines can be adapted to this design, the cost could drop to 70 to 80 cents per ton-mile." He concludes by stating, "there is no doubt that the helicopter with its outstanding characteristics could become one of the most useful and valuable instruments in the hands of progressive lumber operators." We will see in the Chapter 7 how Mr. Sikorsky's prophesy for new logging ventures has come true and is a major helicopter business today using the Crane.

During a May 1960 lecture at the Massachusetts Institute of Technology, Mr. Sikorsky reviewed his efforts over 50 years to conceive of and design helicopters. The Crane helicopter is mentioned at the end of his talk, when he refers to the S-60 Experimental Crane, which "demonstrated a whole new field of extremely useful and valuable services for military as well as commercial uses" as shown in Fig. 1.10. He refers to the S-60 as "a Crane with a 6-ton payload capacity now under construction in our plant." This talk confirmed what he had forecast in his London lecture in 1955, five years earlier.

He then comments,

> In general, the Crane will open up a new and vast field of applications. It could be used in connection with prospecting, mining, oil drilling, installing of electric lines, delivering ready-made homes, and an unlimited variety of other interesting services. In many cases it would promote the growth of new enterprises and industries.

[§]The quotes of Mr. Sikorsky and pictures in this section are found in documents of the Igor I. Sikorsky Archives, Inc.

a) b)

Fig. 1.10 Sketch of Crane: a) building a pontoon bridge, and b) moving a truck from ship to shore.

This vision calls to mind Mr. Sikorsky's comment about the early days of aviation: "Aeronautics was neither an industry nor a science. Both were yet to come. It was an art and I might say a passion." These words apply to his earliest work and also to the Crane. Without his passion (Fig. 1.11) it is clear there would never have been a Crane.

Fig. 1.11 Sikorsky patiently explains another of his new ideas.

CONCLUSION

It was three years before he obtained approval, in 1958, to begin the design of the low-cost, experimental S-60 Crane. The following chapters will explain how his Crane concept was developed as an aircraft and as a vehicle for new military and commercial uses. These chapters will explain why his Cranes continue to be employed today by commercial operators for logging, construction, and firefighting worldwide, 50 years after the design of the first production Crane began.

Early in this chapter, mention was made of Sikorsky's arriving in America with little money. He described this period as follows:

> I have been hungry in America, I have known what it is to seek for work and not find it in America. But there was never a day, during the hardest times, that I lost hope in my planes or that I did not say aloud, Thank God I am here, a free man breathing free air. No man can order what I do. If I fail I can try again.

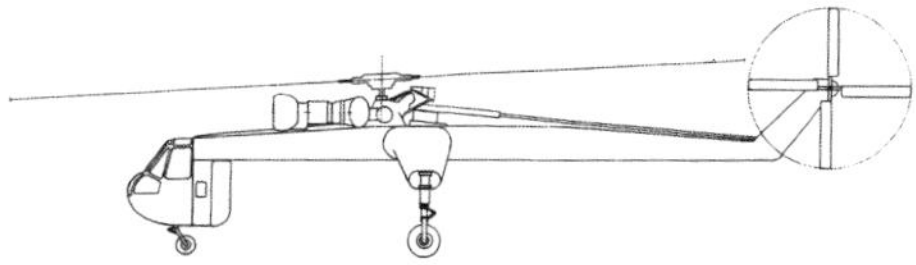

Chapter 2

S-60 Experimental Crane Proves the Concept

S-60 Early Design Work

Lewis Knapp, chief of preliminary design at Sikorsky, recalls his work with Mr. Sikorsky, in early 1958, on the design of the S-60 experimental Crane as follows:

> The Crane had always been on Mr. Sikorsky's mind, since the early 1950s, as the next big step in helicopter history. But there was little interest at the company, because it was very busy designing and producing several types of helicopters at that time. In early 1959, Mr. Sikorsky asked me to sketch a Crane by using production components available at the company. All my other preliminary design engineers were tied up on other development efforts, so I undertook work on his Crane project myself.
>
> As he visited me every day, the design rapidly evolved. It was to create a new Crane type helicopter by modifying the Sikorsky model S-56/HR2S1/H-37 [the large, twin-engine U.S. Marine and Army troop transport helicopter then in production]. The Crane would be a stick fuselage machine capable of picking up from the ground an external load of any size [within the lifting capacity of the crane], of carrying an attached cargo pod, or of moving people on a people platform or in a people pod. As we talked, I understood he had been thinking on this very subject for many years.
>
> As we poured over the drafting board, he would occasionally offer a suggestion. "Lewie don't you think that we might . . .?" If I countered, "but, Mr. Sikorsky, if we did that we might have a problem with . . . ." He would answer "You are absolutely right" and we would go onto the next subject. The following day I might hear, as though it were a new subject, the very same question "Lewie don't you think . . .?" It finally occurred to me that this polite and gentle man could not give a direct order. Quiet persistence was his only way to get what he knew was right.
>
> We replaced the S-56 passenger cabin with a boom fuselage connecting the cockpit with the tail. We used everything we could from the production S-56, which saved time and cost.

The S-56 was the largest cabin helicopter then in production in the United States, five times the weight of its predecessor helicopter. The S-60 sketch in

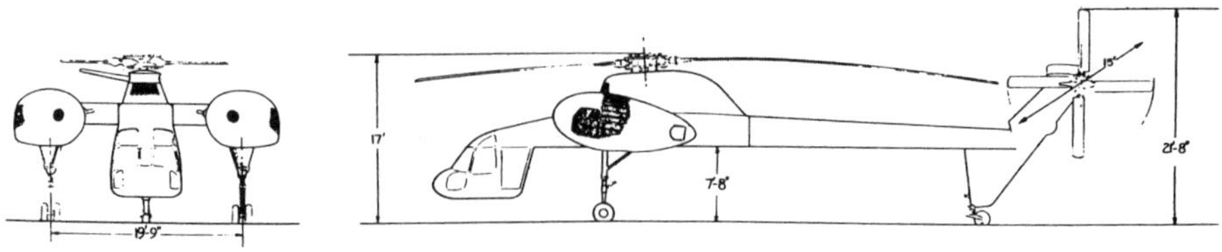

Fig. 2.1 Early sketch of the S-60.

Fig. 2.1 shows the Pratt & Whitney reciprocating engines mounted outside the stick fuselage because they were so large. Again, Lewis Knapp recalls,

> One major difference between the cabin helicopters and the Crane was the location of the cockpit. The Crane cockpit should be below the boom fuselage so the aft-facing pilot can see the load and control the Crane while picking up and dropping loads, but the aft-facing pilot is then below the normal aircraft center of roll. In cabin helicopters the pilot sees both left yaw and left downward roll as a leftward movement, but from the lower cockpit in the Crane, the pilot sees left downward roll as a rightward movement. We called in the experimental test pilots to discuss this and decided to continue as we were. During flight test, we found that, with pilot training, the differences were insignificant.

The exposed cockpit position below the fuselage that Knapp and Sikorsky developed can be seen in Fig. 2.2. Knapp reveals,

> Unlike cabin helicopters, the cockpit of the Crane is not a straight extension of the basic fuselage, but is suspended in space forward and below the stick body structure, called the boom. Thus, if the main landing gear should collapse in a severe crash, the cockpit would be first to strike the ground.

Fig. 2.2 Early flight test of the S-60 at Sikorsky plant in 1959.

> To protect the cockpit from taking the full load of the Crane in such a severe crash, we designed the fuselage, where it is attached to the cockpit, to act as a hinge. In the event of a severe crash, this allowed the cockpit to swivel up into space forward of the fuselage and the fuel cell to improve survivability.

When the S-60 testing was complete, the helicopter crashed while simulating a vertical takeoff and landing (VTOL) fighter takeoff under a NASA contract. It was destroyed, but the cockpit separated as planned and the pilots literally walked away. The accident was not related to any Crane testing.

Approval to Build S-60 with Company Funds and DOD Components

Knapp adds, "Some months later, in May 1958, Mr. Sikorsky obtained approval from the corporation to build a single experimental Crane using company funds, provided we could obtain the S-56 production components for the Crane from the Army or Marines."

The Crane concept now found support from some military leaders in West Germany and the United States. Some believed their armies needed larger helicopters to move larger and heavier loads more rapidly to remote areas and their navies needed to move heavy loads quickly from ship to shore, where there were crowded ports or no ports.

The Crane could move all types of loads (Fig. 2.3) quickly because its wide landing gear allowed it to taxi directly over loads, or hover precisely over loads, as controlled by the aft-facing pilot. The Crane would also have a powerful winch that would immediately pull the load off the ground and position the load at the correct distance from the helicopter for streaming the

a) **b)**

Fig. 2.3 S-60 delivers a) missile load and b) fuel drums.

load behind the Crane for efficient forward flight. The main winch and hook were installed in the S-60 fuselage directly under the main rotor to reduce the movements caused by swinging loads.

When a passenger- or cabin-type helicopter picks up an external load, the pilot cannot see the load once he flies over it, and so he normally relies on signals from ground personnel. One person stands aside from the load so that the pilot can see his hand signals. The pilot moves the hovering helicopter sideways or backwards to get as close as possible to the load. He then moves the helicopter up or down to position the hook, which is attached to fixed cables. Other personnel attach the hook from the helicopter to the load. Often several passes are needed from the helicopter to get a successful hookup. Some helicopters use mirrors to assist these maneuvers. When the load is connected, hand signals are given to the pilot to take off.

Because the Crane could greatly accelerate the movement of material with fewer personnel, the military was interested. As a result, Sikorsky was able to obtain from production all of the S-56 material needed. Final management approval was then obtained to fund the design, building, and testing of this radical aircraft to be designated the S-60. Some of the S-56 components were of older technology, but the S-60 would be a low-cost, low-risk, and rapid way to test Sikorsky's Crane concept.

There were many questions raised by the Crane's design. Would the aft-facing pilot help to move the loads rapidly meet the needs of the military and the commercial users? Would the aft-facing pilot be able to control the aircraft properly and safely in the loading and unloading phase? Would it be practical to taxi over a load, squat over it by reducing the stroke of the landing gear, and attach the load quickly to the aircraft through multiple connections? These and many other questions would be answered by the flight tests of the S-60.

The detail design and construction began in May 1958. It was completed in ten months—record time to design and build a helicopter of this size. The use of Marine Corps components from S-56 production was a major factor. First flight was successful on 25 March 1959, seven days after shop completion, another record. One reason these records were possible was the high regard that Sikorsky employees had for Mr. Sikorsky and their desire to make his Crane concept successful.

S-60 PROVES CRANE CONCEPT WITH FLIGHT DEMONSTRATIONS

The following month, the S-60 also performed flight demonstrations at the Sikorsky plant for a West German military group, which was studying a possible military requirement for a heavy lift helicopter for the German Armed Forces. The S-60 then made a series of flight demonstrations at U.S. Army, Navy, Marine, and other sites in 1959 and 1960, where it successfully demonstrated the wide range of Crane missions.

Fig. 2.4 Igor Sikorsky conducting his daily post-flight examination of the S-60.

Flight Testing on the Open Platform

Mr. Sikorsky continued to be actively involved in the S-60 by examing the records in the cockpit after most every flight, as shown in Fig. 2.4. He also rode on a test flight as seen in Fig. 2.5. S-60 Project Engineer Rod Smith shared the story behind the photographs with Lew Knapp, which Lew recalls here:

> One of the first test items flown on the S-60 was a simple, flat wooden platform built to carry cargo. It was hung on cables under the S-60. With

a) b)

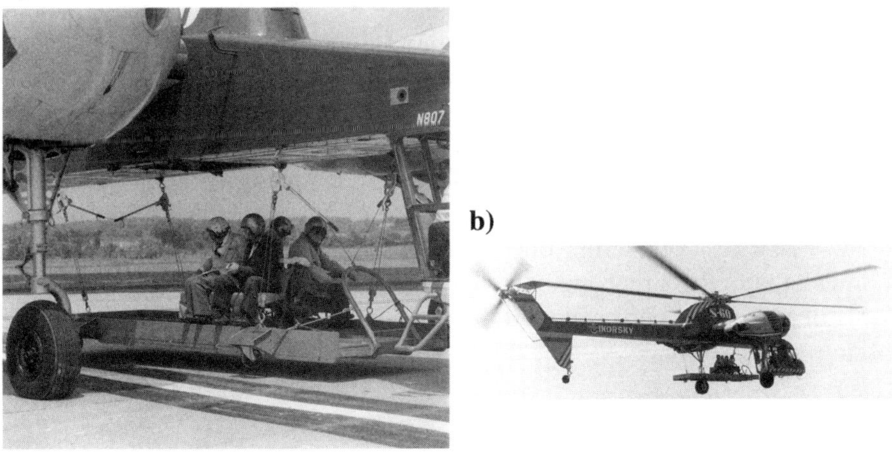

Fig. 2.5 Igor Sikorsky and the other passengers on an open platform, hung under the S-60 on cables, a) strap into their seats and b) enjoy their open-air flight.

> airline chairs bolted on the open platform, it could carry people. In July 1959, Rod Smith, Mr. Sikorsky, Charles Echeverria (project engineer), and John Vest (engineering manager) carefully got on those seats on the platform. They attached their seatbelts and huddled together in their breezy seats.
>
> When they were airborne and flying at 50 knots, they saw some rustling in one of the seats. It was Mr. Sikorsky, who stood up, stepped forward and began to walk around the open platform. As Rod told me, he and the others scrunched down in their seats with faces blanched and bunched together. Mr. Sikorsky walked to one edge, looked down, then strode to the other side. Satisfied, he casually walked to his seat and sat down. Was he researching some question in his mind or was he reliving his flights across the steppes of Russia, many years earlier, on the open deck of his Le Grande the world's first four engine aircraft?

This flight was also described by Roger Dove in his article entitled "The Dime That Didn't Bounce" (probably published in the United Technologies corporate publication *The BEE-HIVE*):

> . . . the flight was amazingly free of vibration. They sensed they were on the threshold of a major discovery. The crane went higher and faster. It finally reached 1500 feet and a speed of 70 knots.
>
> John Vest, the engineering manager, reached into his pocket and placed a dime on the one of the metal floor plates. Vest said it laid so still I could read the date on it. We knew then this was a major discovery.

Very low vibration could be valuable for missions, involving the flight of sensitive military equipment and passengers.

This was not the first time Mr. Sikorsky flew on an open platform hung on cables under a helicopter. When the author interviewed for a job in engineering in 1956, he saw an older helicopter hovering with a simple wooden platform suspended on four cables. As the helicopter made wide turns, the platform flew even wider and away from under the helicopter on the cables. A man was strapped into a chair on this suspended platform, so his guide was asked if that was a flight test engineer on the platform. He was told, "No that is Mr. Sikorsky conducting one of his tests." The author decided, at that moment, that if this great engineer, the inventor and founder of a new industry was still personally performing flight tests, that he would take any reasonable job to work in Sikorsky's company.

PASSENGER POD

Mr. Sikorsky also had high hopes for the use of passenger pods to be attached to the Crane to take advantage of the low vibration level that the Crane had proven it could provide. Engineering designed a simple pod to confirm this, but there were no funds to build it. He was obviously disappointed.

When he came back to work after the two-week company summer vacation, he found the pod was almost completed by shop workers, who decided to give up part of their vacation to build it. It was subsequently flown with good results (Fig. 2.6). The value of the pod concept was confirmed when the U.S. Army bought many pods, which were used by Army Cranes to lift 87 fully equipped soldiers in Vietnam and for other missions.

During the years of flight tests, the capabilities and limits of the S-60 were fully developed. It picked up and delivered a wide variety of large and heavy loads for a wide range of potential users both military and commercial. The ease of pilot training and speed with which loads could be delivered and removed were also proven. All of these data and experience were planned for the possible design of an all new, larger, more capable Crane.

Recommended Changes for the Next Crane

The following changes were recommended by Sikorsky's engineering department for incorporation in the next Crane:

1) Installing a side-arm control to solve the problem of the rudder pedals and azimuth stick interfering with aft-pilot visibility

2) Increasing kneeling capability of main landing gear (thus allowing the aircraft to pick up cargo secured to its hard points off the ground)

3) Adding additional cables and hard points to eliminate swaying of the cargo in flight

4) Building vibration isolation in each hard point to reduce vibration on all loads

5) Adding the ability to jettison all loads

6) Adding a variable-speed winch to allow for location of the load at any distance from the helicopter so that it rides best in the helicopter slipstream

Fig. 2.6 S-60 flying with first pod.

7) Replacing the tail wheel with a nose wheel to assist in landing on or over loads and for towing objects on land or water

8) Adding shade roller devices (later called *load levelers*) to hard points, which would retract when not in use and snug up and lock when secured to a load (thus allowing the use of kneeling to pick up the cargo so that the Crane could pick up cargo, taxi, and make regular or emergency landings)

The conclusion of the engineering report on the S-60 states:

> . . . that the Crane concept, as demonstrated by the S-60, is valid and practical. It can carry any shape or size load within the weight limitations of the S-60 [missiles, pods, vehicles, bridges, water buckets]. It can discharge and pick up loads faster and safer than a cabin helicopter or even a truck. [60 seconds to discharge one load, take off, land over a second load, hook up the second load and take off].

CONCLUSION

Mr. Sikorsky often said ". . . the most dangerous forecast in aviation is to predict the impossibility of something." Perhaps critics who forecast that the S-60 would not greatly accelerate the moving of loads should have heeded this advice from Mr. Sikorsky.

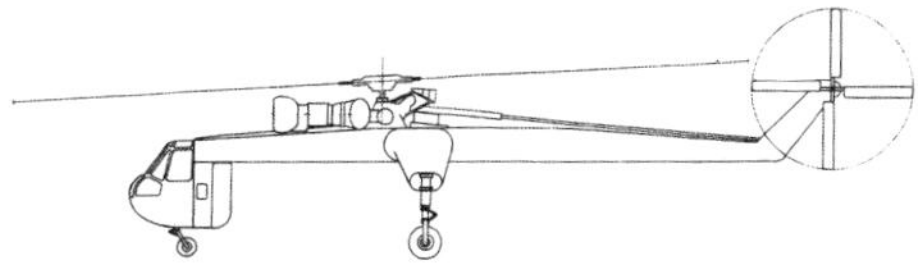

Chapter 3

DEVELOPMENT OF THE S-64 PRODUCTION SKYCRANE

APPROVAL TO BUILD AN ALL-NEW CRANE

The flight tests and demonstrations of the S-60 proved that a Crane helicopter met all of Mr. Sikorsky's goals. Lew Knapp reported that Mr. Sikorsky worked diligently to convince Sikorsky Division management and finally United Aircraft Corporation management to make a much bigger investment in a new, more powerful Crane to be called the Skycrane (the S-64's commercial name), even though there were no orders promised. When Mr. Sikorsky obtained these approvals in April 1961, he approached Lew Knapp, as he had for the S-60 Crane design, to begin design work on a new Crane that would create new military and commercial applications.

As Chief of Preliminary Design, Lew said, "I was heavily involved in the design of other helicopters at that time, so I assigned this task to one of my senior designers, George Howard." This was an opportunity to include the valuable lessons learned from the S-60 experimental Crane in an all-new helicopter. Once Pratt & Whitney defined the new JFTD12 gas turbine engine, the S-64 general design was fixed, and the detail design of the main gearbox (MGB), main rotor, and many other components began.

MAJOR PHYSICAL CHANGES FROM THE S-60 CRANE

The following physical changes defined the S-64 Crane:

1) Replaced the two very large and older piston engines with two newer, smaller, lighter, and more powerful gas turbines, increasing horsepower per engine from 2100 to 4500 shp

2) Designed a new MGB to handle double the power

3) Added a sixth blade to the S-60's 72-foot-diameter main rotor to withstand the higher power (the new blades also had increased chord and other improvements)

4) Added a nose wheel in place of the S-60's tail gear, which had delayed rapid movement of loads under the parked S-60 Crane and delayed taxiing of

the S-60 over loads on the ground (it also impeded the towing of ships or vehicles)

5) Redesigned the aft-pilot cockpit to include additional instruments and controls to allow precise control of the aircraft for quicker access to and delivery of loads

6) Increased cockpit size to include a pilot, copilot, aft pilot, and two mechanics (the Crane often flew more than ten hours per day in remote areas, increasing the need for onsite maintenance)

7) Increased the power of the hoist to carry loads as heavy as 20,000 pounds from 100 feet below the hovering S-64 (the hoist would be mounted inside the fuselage below the MGB, at the center of gravity of the Crane)

8) Added a new system called *load levelers* with the ability to quickly attach to a load [this system consisted of four small winches each capable of lifting 5000 pounds; these were mounted, two on each side of the boom (fuselage), at about 8 feet forward and 8 feet aft of the main rotor shaft]

9) Added a battery-started auxiliary power unit so that the S-64 could start itself anywhere

10) Eliminated all of the cowlings and work platforms on the upper deck of the fuselage

Lew Knapp explains,

> George Howard made another major change, which was the elimination of all the cowlings for the engines, main gearbox, hydraulic stand, accessory box, auxiliary power unit, and tail driveshaft. This reduced the weight, cost, and most importantly, maintenance time. Many of the electric and hydraulic lines and other devices were also outside the fuselage instead of inside, which reduced maintenance even further.

This change gave the S-64 a unique appearance because no other large helicopter had so many major and minor components exposed. The change definitely reduced maintenance time, which helped maintenance personnel, pilots, and aircraft managers, because S-64s were available to return to flight status more rapidly. The increase in drag was small, especially in the high percentage of crane missions of 5 to 20 nautical miles at 40 to 80 knots.

The ready access to the many major components and critical smaller parts mounted on top of the stick fuselage was an especially valuable benefit for commercial Cranes operating up to eleven hours per day. Routine and emergency maintenance was often performed on commercial Cranes with temporary lighting at night in remote sites. Clearing 9 feet 4 inches from the ground, the first S-64 Skycrane is pictured in Fig. 3.1. The flat underside of the fuselage is free to attach large, wide, or long loads. Note the retracted tail skid to stay clear of loads. The fuselage top provides a flat surface to speed inspections and repairs of exposed components.

Fig. 3.1 First S-64 Skycrane.

UNIQUE DESIGN FEATURES OF THE S-64

The S-64 was planned for production with the most innovative design features of any helicopter. It is important to summarize them to fully understand how radical a design was the Crane.

NO CABIN

The absence of a cabin provided a ground clearance of 9 feet, which made possible the rapid attachment of external loads driven directly under or towed under the Crane. With the load attached to the underside of the Crane fuselage, drag was reduced on longer flights. This design also reduced weight and cost by eliminating the cabin with its floors, doors, ramps, windows, seats, and other components. Having no cabin also increased payload by over 1000 pounds.

SIMPLE STICK FUSELAGE

A simple stick fuselage provided a flat top deck to more efficiently mount engines, MGB, hydraulic control panel, auxiliary power unit (APU), tail driveshaft, and other critical components. This simple arrangement gave quicker access to almost all key components.

NO TOP DECK ENCLOSURES

No top deck enclosures on engines, MGB, and so on, saved several hundred pounds of weight and reduced maintenance time. The small increase in drag was not a significant penalty when a high percentage of missions were 5 to 20 miles at low speeds.

FOUR LOAD-CARRYING METHODS

No helicopter has had this many options for moving cargo:

1) The main hoist lifts any size load up to 20,000 pounds at 100 feet below and draws it up into streaming position behind the Crane for best cruise speed. As the Crane comes to a hover, the hoist is used to position and release the load.

2) Four small hoists mounted on the side of the stick fuselage 8 feet forward and aft of the centerline of the main rotor can lift 5000 pounds each to pick up equipment, pallets, or other loads. These hoists can draw the load up against the fuselage for longer flights with reduced drag.

3) For odd-shaped loads or those with no lift points, like a collection of lumber, which was a frequent load in Vietnam, there are 32 hard points on the bottom of the fuselage that are used to secure these loads.

4) The pod can be quickly attached with the four hoists. The pod is large enough to seat 87 fully equipped troops. When equipped as an operating room, the pod has carried surgeons directly to the edge of the battlefield to save the lives of the severely wounded. Pods have also served as command posts, maintenance shops, first aid units, and air traffic control sites. The Crane can fly to the location of the pod, pick it up in a few minutes, and deliver it quickly to where it is needed.

KNEELING LANDING GEAR

The kneeling landing gear accelerates the attachment of a vehicle, a pod, or other loads moved under the standing Crane. Kneeling of the Crane moves the load closer to the bottom of the stick fuselage for quicker engagement.

AFT-FACING PILOT IN HIS OWN COCKPIT WITH HIS OWN FLIGHT CONTROLS

This arrangement was a first for a production-helicopter flight-control configuration. The aft pilot flies backward except when he directs the aircraft to fly rearwards, which he often does to pick up or deliver loads. His collective control is mechanically linked to the pilot's collective control (which has override authority), and his side-arm control provides limited cyclic and yaw control.

Controls for the main hoist and the four small hoists are provided as are the jettison controls for these hoists. Total engine output torque, hoist cable length, and load weight have indicators as well. The pilot has an excellent view of the load and the ability to control the aircraft to precisely deliver the load within inches or position the hoist so that it can engage a load quickly. These controls made it possible later, for example, to move cargo from a ship underway.

ELECTRIC SIDE-ARM CONTROL FOR THE AFT-FACING PILOT

This control was required by the small size of the cockpit and was believed to be the first time a nonmechanical flight control was used in a production helicopter. Pulling back or pushing down on this control moves the nose of the Skycrane up or down. Moving the control left or right moves the nose left or right. This control made possible rapid and precise movement of the helicopter and the load for rapid and precise placement. The control could be overridden by the pilot.

MISSIONS OF THE S-64

These unique design features combined to make it possible for the Skycrane to perform many military and commercial missions faster than any other helicopter. They also allowed the Skycrane to perform missions that no other helicopter could perform.

Table 3.1 shows a remarkable increase in payload for the S-64E over the S-60. The first S-64 built was later designated S-64E, when it obtained full Federal Aviation Administration (FAA) certification. Army CH-54A and CH-54B performance is similar to the FAA-certified S-64E and S-64F before the addition of armor plate for pilots and engines plus other Army requirements. When a more powerful engine became available, the Army funded an improvement program to develop the CH-54B, which was produced later in the Army production run.

The engine horsepower of the S-64E is 114% higher than the S-60, and the engines are lighter. The empty weight of the S-64E is about 400 pounds less than the S-60, even though it is larger, has an extra blade, a larger rotor head,

TABLE 3.1 COMPARISON OF S-60 AND S-64S WEIGHTS AND PERFORMANCE[a]

	S-60	S-64E[b]	S-64F[b]
Power (30 min.)	4200	9000	9600
Main rotor diameter, ft	72	72	72
Number of main rotor blades	5	6	6
Weight empty, lb[c]	19,600	19,200	19,700
Payload, tons	6	10	12.5
Payload increase over prior model, %	—	67	25
Gross weight, lb[c]	34,500	42,000	47,000
Cruise speed, kn	100	95	105
First flight	1959	1962	1968

[a]All comparisons are for commercial versions of these Cranes because there was no military version of the S-60.
[b]S-64 blades had wider chord and other improvements to generate more lift.
[c]Weights are rounded to the nearest 100 lb for ease of comparison.

a larger MGB, and other additions. The S-64E gross weight is only 22% higher than the S-60, but the payload is 67% higher, primarily due to the higher engine power, lower engine weight, an added blade, and other improvements.

These differences are due to the S-60 being a low-cost, experimental Crane that used older engines and other components designed many years earlier. The S-60 goal was to quickly prove the concept. The S-64E was an all-new Crane using the latest technologies and lessons learned from the S-60. The cost to design, build, and test the S-64E was much higher, but so was the performance.

PRODUCTIVITY OF CRANE- VS CABIN-LOADED HELICOPTERS

The Crane pickup time on most loads with the hoist rarely exceeds 30 seconds and is often less. The Crane can deliver loads in less than 15 seconds, once it is in a hover. When the Cabin is in a hover, it must then land, open its ramp, wait for a vehicle to bring the load, move it inside, calculate the new center of gravity, reposition the load, close the ramp, and take off. In Vietnam, this usually took 15 to 20 minutes. The Cabin can fly faster with the internal load, but with 86% of missions at less than 20 miles, according to one Vietnam study, this did not save much time. Most commercial Skycrane flights are also less than 20 miles. In addition to lower productivity, the Cabin has more than 30 times the exposure to enemy fire, which often occurred in Vietnam in these missions.

PRODUCTIVITY OF THE CRANE VS EXTERNALLY LOADED CABIN HELICOPTERS

The Cabin can pick up loads externally, usually with a two-hook, fixed sling. It will take more time for the Cabin to position itself over the load, using mirrors and signals from ground personnel. It will take about the same time as the Crane to deliver the load, but the weight of the load will be less than the Crane for the same installed power. This is because of the lower empty weight of the Crane.

Only the Crane can deliver external loads precisely in a few minutes without the help of personnel, such as a new bridge section ready to be bolted in place, or a radar to the top of its tower, or a section of a new power line tower on top of other sections. The Crane can also reach down 100 feet to retrieve a downed helicopter or deliver equipment or supplies to troops deep in the jungle or in a narrow ravine.

The size of the cargo is not a problem for the Crane; for example, it has moved towers over 90 feet in length. The Crane can also taxi over a load, pick it up on four load levelers, and make a rapid takeoff. The S-64 was able to do this in 60 seconds, which is faster than loading a truck, in one engineering

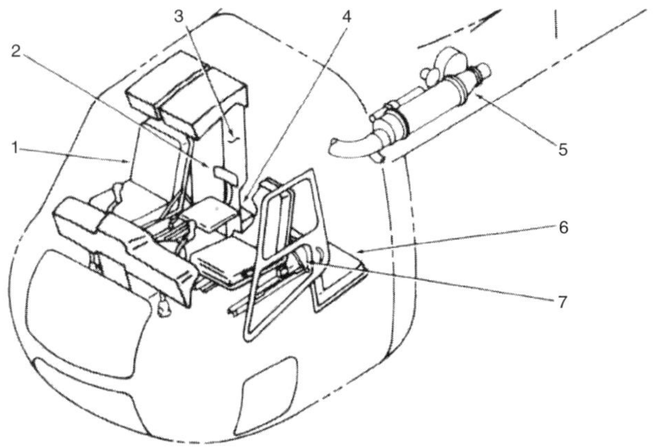

Fig. 3.2 Arrow 6 shows the aft-facing pilot seat behind and below the copilot's seat.

test. Time is critical in both military and commercial operations; the Crane design has proven to be the quickest, aided by the visibility afforded by the aft cockpit position shown in Fig. 3.2.

Basic Power and Control Systems of the S-64

The MGB drives the main and tail rotors, two electric generators, and four hydraulic pumps. The main rotor diameter is 72 feet and has six blades, fully articulated. The tail rotor diameter is 16 feet with four blades and is semirigid. Pilot inputs to the rotors are amplified and modulated by an automatic flight control system using servos. Electric power is a three-phase, 400-cycle, 115/200-volt ac system with a backup 28-dc system.

Crane Mode vs Cargo Mode of Lifting

From the pilot's viewpoint, the Skycrane flew in two different modes. In the Crane mode, loads are those carried on the main hoist and are often too large to attach to the stick fuselage. Smaller loads are also carried in the Crane mode when initiated with short notice (often with the military). Such loads have little planning, and so the payload, weight, center of gravity, and aerodynamic and dynamic behavior are not accurately known. These conditions require that the means of attachment be fast, simple, and reliable. Load release and emergency jettison ability while airborne is necessary and is provided.

In the Cargo mode, as shown in Fig. 3.3, the load is connected to the S-64 by four short cables or 32 hard points. The load never extends below the landing gear. The load must meet dimensional constraints, and the weight and center of gravity of the load must be known with reasonable accuracy because the load is so close to the fuselage and affects aircraft flight handling.

Fig. 3.3 The truck is attached to the fuselage by four load levelers in the Cargo mode.

The Crane can reach higher speeds in this mode, and so it is useful for longer range flights.

This description illustrates why loading and unloading times are important to productivity for short-haul loads. Data from actual CH-54 Flying Crane combat experience, provided by (Lieutenant Colonel, Retired) Gary R. Heffner, former commander of the 273D Assault Support Helicopter Company (Heavy) in Vietnam, found that 86% of flights were 20 miles or less. CH-54 hover time for picking up 73% of the loads required less than 15 seconds. Ninety-five percent of load deliveries were made in less than 15 seconds. These actual data confirm that the Crane configuration has high productivity as just explained.

In commercial logging operations, an S-64 delivered 16,000-pound bundles of freshly cut logs from a mountainside to a truck over one mile away and 1000 feet below. The S-64 flew these loads up to twelve hours per day, less the time for several hot refuelings. The time from pickup to delivery to return for the next was usually in less than four minutes. This is another example of the high productivity of the Skycrane.

DEVELOPMENT OF THE FIRST S-64

The detailed design of the S-64 started in April 1961 with the first flight coming in May 1962, 13 months later. The main reasons for this short design and construction cycle were the simplicity of the fuselage vs a complex cabin fuselage and the design concepts proven by the S-60. The decision to eliminate the large number of cowlings and the mechanisms needed to open and support them so they could become work platforms also saved many design hours.

Fig. 3.4 8000-horsepower main rotor test tower (notice man at base of tower).

As key components were fabricated, they were subjected to an exhaustive series of inspections and tests. The major components were tested for performance and endurance in special test stands. The main rotor and blades were tested at various speeds and load levels on the main rotor test tower, as shown in Fig. 3.4. The tail rotor was also tested on a separate stand as were the three gearboxes.

At the same time, all of the major elements of the helicopter drive system were assembled on the custom-made steel structure shown in Fig. 3.5, called the propulsion system test stand, which was anchored in a large concrete pad. This included the two engines, three gearboxes and related driveshafts, two rotors, and several control systems. Power levels simulated aircraft operating conditions up to 125% of flight loads and also included endurance testing. Many design changes came from this test stand.

The fuel control, for example, had to be modified repeatedly to adapt to the dynamic inputs of the many rotating components (two engines, two rotors, three gearboxes, multiple driveshafts, etc.) as different flight modes were simulated. Lee Jacobson, retired Sikorsky director of Aircraft Design and Development, explained that a general helicopter dynamic computer analysis (GENHEL) was developed at Sikorsky to integrate the dynamics of these rotating components. This helped define the engine fuel control dynamic characteristics, which was one of the last steps of this test program. GENHEL is a major design tool that is still in use today.

Fig. 3.5 S-64 propulsion system test stand.

The first S-64 was completed as the drive system and many other components were well along in testing. The aircraft then went through ground tie-down testing with the S-64 attached to cables secured to another massive concrete pad. The aircraft was run at various power levels to verify functions before first flight.

The first flight, on 9 May 1962, was followed by accelerated flight tests. Al Pariaro, the crew chief on the first S-64, reported that Mr. Sikorsky often inspected the S-64 at the end of the day. Pariaro relates, "Mr. Sikorsky usually did a walk-around inspection of the aircraft, climbed into the cockpit, pulled out his pad and took notes for as much as an hour." He surely was tracking the progress of his creation, as he probably had done for all of his aircraft.

First S-64 Flies to Army Base Two Months after First Flight

The S-64 quickly met forecasted flight criteria and reliability requirements, so it was scheduled for demonstrations at several Army bases. In July 1962, just two months after first flight, the S-64 made its first cross-country flight to Fort Bragg Army Base to perform flight tests with standard Amy loads. This was a very short time after first flight for this all-new, large helicopter of radical design to fly off to demonstrate to the key customer. The S-64 lifts a 155-mm howitzer in Fig. 3.6.

Fig. 3.6 Lift of 155-mm howitzer with power pack at Army base.

ADDITIONAL TESTS AND DEMONSTRATIONS

After flight demonstrations at several other Army bases, the S-64 returned to Sikorsky Aircraft to finish flight testing. It also continued to perform many other off-site demonstrations to Army, Navy, Marine, and commercial prospects. A demonstration at the Sikorsky plant to a group of West German

Fig. 3.7 Alex Sperber (left) and the author, Jack McKenna, inspect an S-64 damaged in a hard landing.

military personnel led to a contract to perform two years of flight tests of the S-64 performance for the German military. Two S-64s were built for this and were shipped to Germany early in 1963.

Most flight testing at Sikorsky was accomplished without difficulty, but one flight test, with a container, made a hard landing in a field near the Sikorsky plant (Fig. 3.7). The tail cone could not support the tail pylon, stabilizer, two gearboxes, and the tail rotor during the high-velocity landing. The failure partly severed that assembly of components and a tail rotor blade hit the ground. The author and his superior, factory manager Alex Sperber, flew to the site in the company S-58 to assess the damage and determine how to get the damaged aircraft back to the plant for repair. The design of the tail cone structure was strengthened so that the S-64 could be repaired and returned to service by the author's experiment shop department.

ARMY ORDERS SIX YCH-54AS

After more than a year of successful demonstrations at several bases, the Army ordered six YCH-54As in June 1963. The first flew in twelve months, and four were delivered four months later by October 1965 to the Army at Fort Benning. After two months of testing by the Army, the four were sent to Vietnam in December 1964, where they were rapidly deployed in combat and logistics support operations. The remaining two were used by the Army to complete its testing and obtain FAA certification in mid-1965 for the CH-54A configuration. Sikorsky obtained FAA certification for the commercial version of the S-64 in 1969.

Fig. 3.8 Igor Sikorsky holds an S-64 model of his creation in his office.

The S-64, with all its innovations, proved that it was a sound design that could be rapidly developed and then quickly integrated into an Army operating unit and sent off to war. The CH-54s made great contributions to the Army effort in Vietnam, as explained in Chapter 5. Commercial S-64s were also demonstrated and sold, as explained in Chapter 6.

CONCLUSION

The S-64 Skycrane would not have existed without Mr. Sikorsky's vision and sustained efforts over many years. This thought brings to mind one of Mr. Sikorsky's often repeated quotes: "The work of the individual still remains the spark that moves mankind ahead." It was certainly his spark that created the S-64 and its success.

In Fig. 3.8, one can see the spark in Mr. Sikorsky's eyes as he holds a model of the S-64 in his office. He was able to follow the flight testing of the S-64 on a daily basis—his office in the engineering department was close to the flight field.

Chapter 4

WEST GERMAN MILITARY VERIFIES S-64 PERFORMANCE

GERMAN NEED FOR HEAVY LIFT

Because NATO airbases were vulnerable to a sudden Soviet attack in the 1960s, the Air Force of West Germany was concerned that a surprise attack could severely damage its F-86 and F-104 fighter forces. Germany had begun development of VTOL fighters that would not need concrete runways and could be hidden in multiple and remote sites.

To be effective, these fighters needed to complete repeated missions. This meant the sites required frequent and rapid deliveries of fuel, weapons, and other critical items to small areas, such as places hidden in the woods. The Sikorsky S-60 Crane helicopter had demonstrated how this need could be met by lifting 6-ton loads to a German military group, which visited the

Fig. 4.1 Large truck is lifted by four load levelers attached to two steel plates seen under the truck.

Fig. 4.2 Smaller truck is attached with load levelers.

Sikorsky plant in 1959. A second German military group visited Sikorsky Aircraft in 1962 to witness the S-64 lift payloads of 10 tons that was close to what was needed.

S-64 WINS COMPETITION

NATO ran a competition for the best helicopter to support these fighter sites, and the VFW-Sikorsky proposal for the S-64 Crane was declared the

Fig. 4.3 A long steel beam is picked up by the main hoist.

Fig. 4.4 Missile and its carriage are lifted by load levelers.

winner. The German Ministry of Defense purchased two S-64s in 1961 and funded a flight-test program to evaluate their performance in a series of German military missions. United Aircraft Corporation (UAC) had authorized its Sikorsky Division to work with the German-based VFW Company, in which UAC had a financial interest. VFW had sent twelve of its engineers to Sikorsky in 1961 to assist in the design of the first S-64. The first S-64 was built and flew in 1962, and two additional S-64s were delivered to Germany by early 1963. The photographs in Figs. 4.1–4.7 are examples of some of the many flight tests performed by the S-64s for the German government.

Fig. 4.5 Four steel beams are strapped together and lifted with cables.

Fig. 4.6 Pod built by VFW to move personnel and equipment to test sites.

VFW TEST PROGRAM SUCCESSFUL, BUT REQUIREMENT CANCELLED

Ulrich Heider, of VFW, managed the German test program, with Sikorsky Aircraft personnel in a support role. Ulrich states,

> The S-64s were successful in meeting the German Air Force and Army requirements in an extensive series of tests lasting over two years. When the planned development of the VTOL fighters was cancelled by the German Ministry of Defense, the S-64 German test program was also cancelled in 1965. The two S-64s were then returned to the Sikorsky plant.

The helicopters were later sold as commercial S-64 Skycranes.

SIKORSKY/VFW TEAM COMPETES FOR NEW REQUIREMENT

At about this time, a German requirement for a large troop transport helicopter was developed to replace the aging and smaller Sikorsky/German S-58/H-34s. The competitors were the Boeing CH-47, the French Super Frelon, and the Sikorsky CH-53. Sergei Sikorsky, who worked at the United Aircraft Corporation office in Germany, prepared for this competition by working with the German government.

ORDER OBTAINED FOR 110 CH-53S

The order was won in 1968 for 110 of the Sikorsky CH-53s, whose main rotor of 72-feet diameter with six blades was similar to the 72-feet diameter and six blades of the S-64. The tail rotors were also similar. Both aircraft also had a gross weight of 42,000 pounds, and so it is believed that the two

a) b) c)

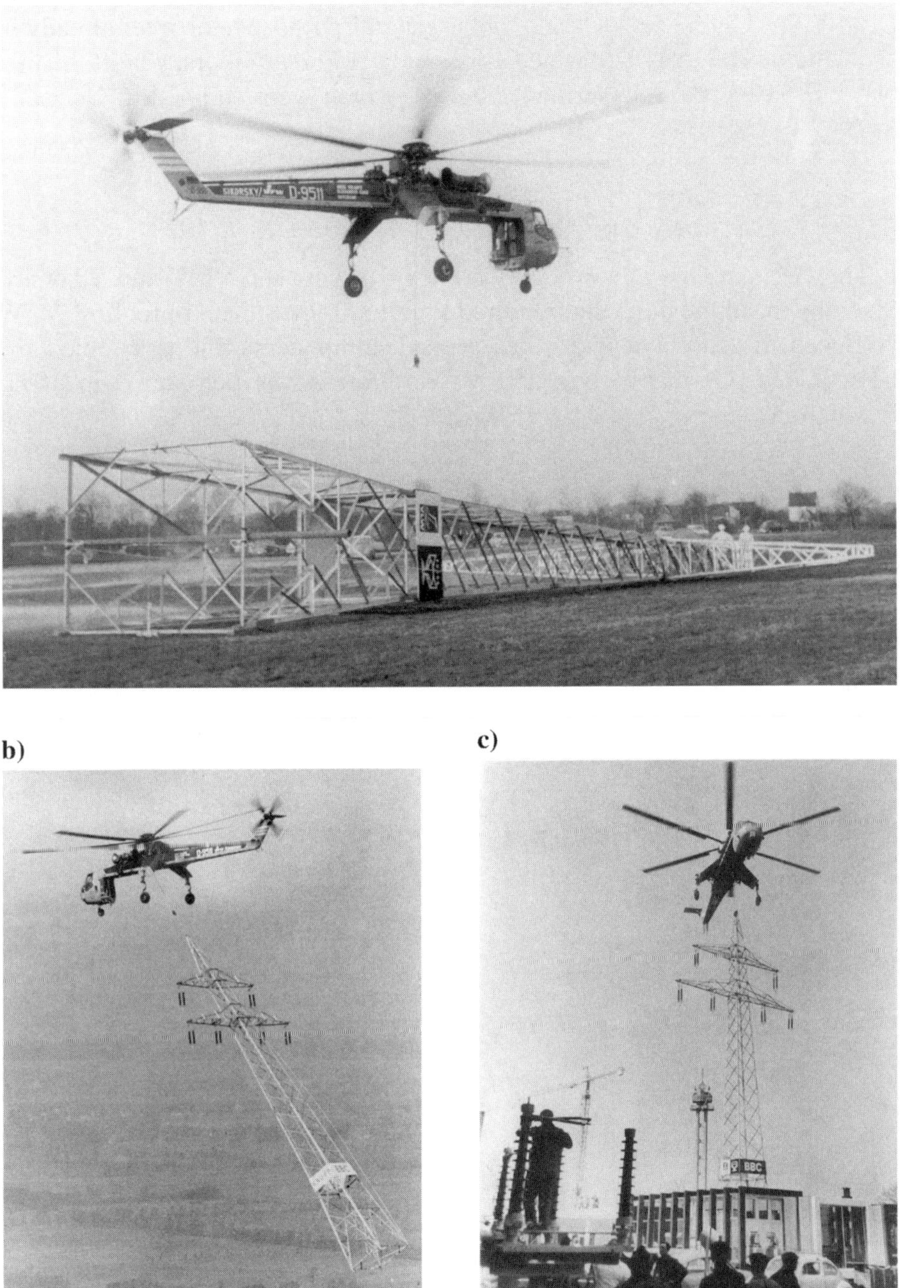

Fig. 4.7 S-64 delivers a 90-foot-long radio tower: a) pick up, b) in flight with tower in streaming position, and c) delivery.

years of S-64 testing helped to obtain a positive decision for the Sikorsky CH-53. The CH-53 was a large-cabin, general-purpose helicopter already in use with the U.S. Navy, Marine Corps, and Air Force. Sikorsky built the first two CH-53G ("G" for German) aircraft, which were shipped to Germany in 1969.

SIKORSKY/VFW COPRODUCTION BEGINS IN 1971

The German CH-53G production was split between VFW and Sikorsky. Sikorsky made the parts that required long-term investment in tooling. VFW produced the fuselage and other critical components and assembled the CH-53Gs in Germany. The first VFW aircraft was completed in 1971. Production was completed by VFW in 1975.

CONCLUSION

With the delivery of the 110th CH-53G, West Germany became the largest helicopter operator in Europe. These helicopters continue, in 2010, to provide air mobility for the German Army within Germany and in other countries. Current plans are that they will continue in service until at least 2030.

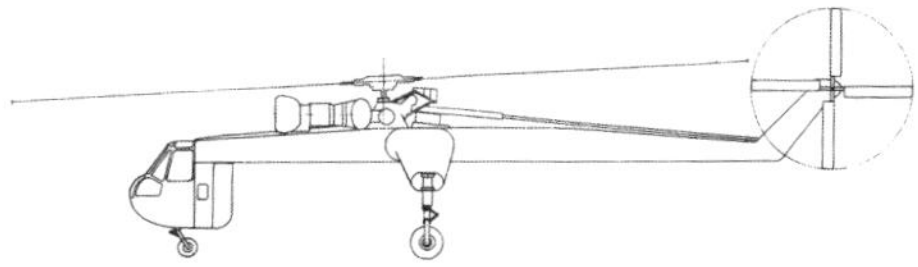

Chapter 5

DEVELOPMENT OF ARMY CRANE AND VIETNAM SUCCESS

KOREAN WAR EXPLOITS HELICOPTER EXTERNAL LIFT

Military needs for a crane-type helicopter were well defined by Peter Brown, Jr., a retired military officer and chief of advanced planning at Sikorsky Aircraft, in a November 1964 issue of the American Helicopter Society's *Vertiflite* publication. His comments were based on the analysis of data generated from cabin-type helicopters used to move both internal and external loads during the Korean War. H-19s like those shown in Fig. 5.1 were deployed in the Korean War.

Brown's summary states that

> Operations in Korea indicated that the most important duty of the helicopter was movement of materials, usually hauled externally. . . . Military operations demand larger and larger payload capability . . . this applies not only to single large loads but to economical movement of many small loads . . . external load handling features could and should be improved. Technology and new subsystems would allow a helicopter to be built with higher payload capabilities and improved load handling characteristics.

a) b)

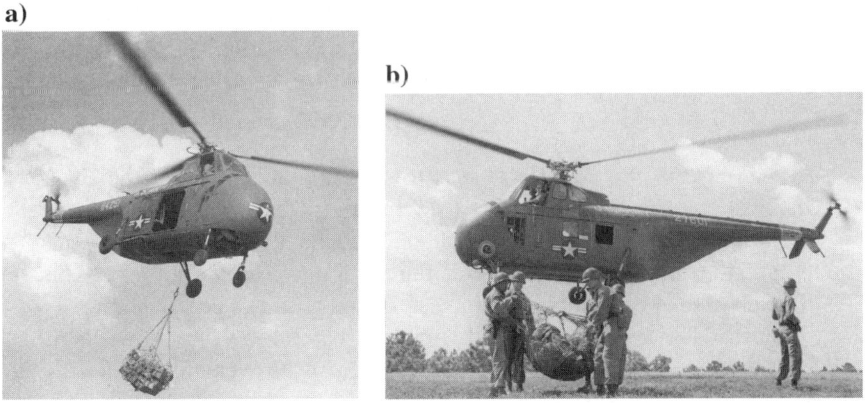

Fig. 5.1 Sikorsky S-55/Army H-19: a) delivers a suspended load and b) another prepares for takeoff.

A Sikorsky Aircraft helicopter history report states that twenty H-19s were assigned to each helicopter transport company in Korea. One of the first combat missions in April 1953 was the delivery of 17 tons of equipment to an 8th Army unit that had been cut off from supplies for two days. The unit was 300 yards forward of the main line of resistance. Fourteen H-19s also supported the 25th Division with 696 flights to carry 563,673 pounds of ammunition, fuel, and food. The H-19 had a payload of just over 2000 pounds compared to the CH-54A payload of 20,000 pounds and the CH-54B payload of 25,000 pounds.

Brown concludes,

> Ton-mile costs are lower with large aircraft.... A large aircraft is less vulnerable than a small one, if maneuverability and speed are not compromised. The tests of the S-60 proved, without question, the worth of the Crane design for faster and more accurate load handling.

The S-64 flight demonstrations at several Army bases in 1962 and 1963, as well as the external load experience of the Korean War, caused the Army to consider the S-64 for its Air Mobile Concept. This concept would allow an Army division to use helicopters as the prime movers of personnel, supplies, equipment, and other cargo.

Army Development of YCH-54s for Vietnam

In June 1963, a $13,000,000 Army fixed-price contract was placed for six YCH-54As. After development and testing, they would be certified as CH-54As. The Y indicated a prototype aircraft that had not been through full flight certification by the Army or the FAA. Jay Rickmeyer was an Army civilian aeronautical engineer assigned to coordinate the YCH-54 Development Project, reporting directly to the program manager, Colonel Denhart. The first two paragraphs of the preface for the YCH-54A Program Master Plan (dated 14 April 1965, near the completion of the program) are reprinted here:

> The purpose of this Master Plan is to schedule all phases of the FAA certification program necessary to achieve a Type Certificate for the S-64A [YCH-54A] heavy lift helicopter. This Plan offers a graphic method to continuously check the progress of this program and would be of value to the FAA in its planning associated with this certification.
>
> The S-64A, a Crane-type helicopter, is basically a new concept in Rotary Wing Design. Its primary use is as a heavy lift vehicle to transport of externally carried jettisonable cargo and its secondary use is with a fixed pod for carrying cargo and equipment. It can carry cargo into areas that are inaccessible to ground vehicles, hover over a ship at sea while

> loading or unloading cargo, tow massive military and commercial ships across open bodies of water or lift almost any object within its heavy lift capacity. Sikorsky had initially built three aircraft, two in service with the [West] German military and one company aircraft for research and development and [for] commercial and military demonstrations. At present, these aircraft and five recently delivered to the U.S. Army have accumulated in excess of 2000 [flight] hours.

CONTRACT GOALS

There were several major and diverse goals in this contract. The first priority was to obtain four YCH-54s for testing with the Army's new Air Mobile Division, as soon as possible. The first aircraft was delivered in June of 1964, twelve months after contract award. Three more were delivered in the next three months. The objective for these four was to learn if this large, unique helicopter would add to the new Army air mobile concept. The fifth helicopter went to another Army base for other important testing, and the sixth remained at Sikorsky to obtain flight certification.

The second priority was to obtain the FAA certification by flying the sixth YCH-54A with the original S-64. The S-64 started the FAA flight-test program nine months before the sixth YCH-54 was delivered to FAA flight testing in November 1964. The earliest date for flight certification was mid-1965, one year after the Army received the first YCH-54 and six months after the four YCH-54s were shipped to Vietnam. It is hard to identify any helicopter that was sent into the combat zone before full flight certification, but there was confidence in the Army, as well as at Sikorsky, and it proved correct.

The entire contract had ambitious schedules, which were aided by its fixed-price nature, instead of the usual cost-plus contract employed for such major engineering and flight developments. Under the contract, Sikorsky was committed to completing each task for the contract price. Additional funds could not be obtained for unexpected problems with planned contract tasks as could be obtained under a cost-plus contract. There were considerable design and development efforts required to change the commercial S-64 design into an Army combat-ready configuration with armor plate, self-sealing fuel cells, military electronics, and more. The FAA test program is shown in Fig. 5.2.

FAA Flight Certification was scheduled for 31 July 1965. Rickmeyer participated in writing specifications, defining requirements, funding procurements, monitoring testing, and FAA certification. Though not flight qualified, in his limited flight opportunities he did hover the crane over a load from the back seat and found "it was very natural and easy to control." He also flew in the pod and found the vibration level to be as low as Mr. Sikorsky had forecast.

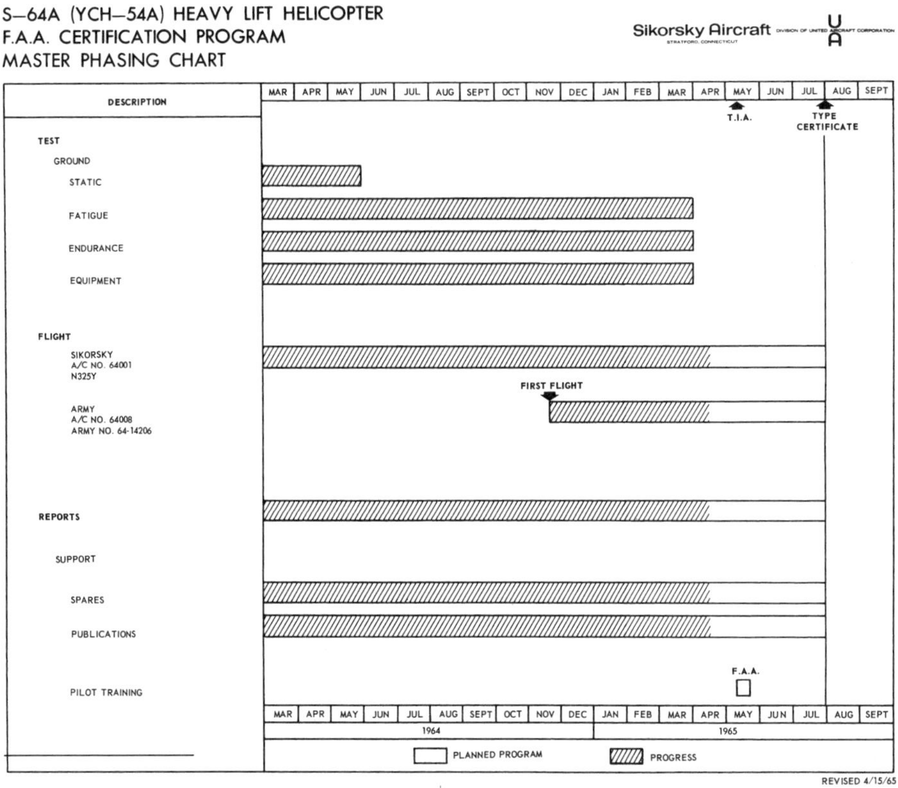

Fig. 5.2 S-64A (YCH-54A) FAA Certification Program master phasing chart dated 15 April 1965 (courtesy of U.S. Army, provided by Jay Rickmeyer).

Rickmeyer reported that he was concerned about the Pratt & Whitney JFTD12 gas turbine engine, which had been developed for the first S-64 without extensive testing. It was derived from the J60 jet engine, which had seen many years of successful service on both Air Force and Navy aircraft. The major change to the J60 jet engine was adding a gas turbine, and its driveshaft connected directly to the MGB of the S-64. A new exhaust stack was designed to divert the hot gases away from the MGB of the S-64.

The plan was to take the JFTD12 engines off the YCH-54s after every 150 hours of flight for inspection. Rickmeyer reported the engines were in such good condition after the first 150 hours that the 150-hour cycle was extended. The engines were up to 600 hours time between overhaul by the ninth month of the flight-test program and proved reliable in Vietnam. An installed engine is shown in Fig. 5.3. The constant section behind the air intake is the jet engine, and the tapered section is the gas turbine, which sends a shaft to the gearbox and finally the exhaust pipe used to divert the hot gases from the MGB. A protective plug is inserted in the exhaust-pipe exit.

Fig. 5.3 JFTD12 engine inlet is on the right (near the cockpit also on the right).

Pilot training at Sikorsky also began very early in the program, and Chief Warrant Officer (retired) R. J. Oden explains that he

> was one of the first three instructor pilots trained on the YCH-54s. Each pilot to be trained was required to have a minimum of 1500 hours of helicopter time, a helicopter instrument rating, and be twin engine helicopter qualified. The selection requirements were this high because everyone from General Kinnard down wanted the YCH-54 to succeed in the new Air Mobile Concept being developed at Fort Benning. It was a joy and pleasure for me to fly with this group.

Rickmeyer also reported that some important design changes came from the test program. A beanie-shaped cover was added to the top of the main rotor to protect instrumentation (Fig. 5.4). When removed, the pilots complained of increased irregular airflow and vibration. The beanie was

Fig. 5.4 Part of the beanie can be seen on top of the main rotor above the engine exhaust; one of the load levelers with its light, yellow hook is in the foreground (courtesy of Raymond L. Robb). Skycrane is facing to the left here, as compared to Fig. 5.3.

reinstalled, which eliminated the vibration. It was added to all Cranes. A split track of the rotor blades was also found to reduce vibration and was added to all Cranes.

Army operations called for more range later, and so a fuel tank was added in the tail cone on the production line. When the troops complained that the tank was impossible to install as a retrofit in the field, a quick visit to the production line by Rickmeyer showed the troops were right. He found a very small worker to crawl through the very small access area in the tail cone for the installation. The access area was enlarged, and the problem was solved.

YCH-54s Go to Fort Benning and Then to Vietnam

The first Crane was delivered to Fort Benning in June 1964 with Sikorsky personnel in a support role, as Army personnel completed training. Lieutenant Colonel (retired) Paul J. Fardink reports in his article in the December 1996 *Army Aviation* magazine: "During these exercises, a detachable pod was converted into a tactical operation center for General Kinnard."

Fig. 5.5 CH-54 delivering ammunition in Vietnam to a 155-mm howitzer crew supporting combat operations.

Rickmeyer continued,

> Two months after arrival of the fourth YCH-54 at Benning, the four Cranes were sent to Vietnam in December 1964 to join the first Air Mobile Division. Upon arrival in Vietnam, about May 1965, the four helped off-load equipment from the carrier in a ship-to-shore operation. Loads were moved, in record time, directly to the 11th Air Assault Division. The Cranes were quickly put in service to deliver artillery and ammunition to combat areas [Fig. 5.5] as well as hauling bulldozers and structures to Forward Fire Bases.

The FAA Type Certificate was obtained in August 1965 so that the four Skycranes supporting combat operations could drop the Y and be designated CH-54As.

A production order was placed in 1965 with deliveries begun in 1967 (Figs. 5.6 and 5.7). Production reached over two per month in 1968 and ended in 1972 with a total of 95 CH-54s delivered to the Army. The CH-54s cost the Army $2,847,303 each, and the CH-54Bs cost $3,014,803 each. Commercial operators have paid as much as $28,000,000 each for a surplus CH-54 in remanufactured and updated condition about 40 years later.

Fig. 5.6 Walter Lysak, Brigadier General Delk Oden, and Igor Sikorsky at a production CH-54A.

Fig. 5.7 CH-54 production line.

ARMY OPERATIONAL MANUAL DESCRIBES CH-54 USE

The Army Operations Manual for the CH-54A states, "The aft pilot is seated back to back to the copilot and faces aft. He is provided with an electric hover control stick (cyclic and yaw), a collective pitch control, and cargo handling instruments. The pilot's compartment seats 5 crew members; the pilot, copilot, aft pilot, and 2 crew members."

The manual defines

> three load carrying methods [1] to carry single point loads . . . a cargo hook is located at the center of gravity [directly below the main rotor], [2] there are 32 structural hard points [on the bottom of the fuselage] which can be utilized to carry loads and [3] the load leveler system consists of four load suspension units, which are raised and lowered by four hydraulic cylinders to produce load leveling capability.

The 32 hard points and the four load levelers were used to snug the loads up against the fuselage for low drag, longer flights.

Another load-carrying method was the pod, which could be configured to carry troops, command centers, medical centers, or other missions. Instead of different helicopters dedicated and configured to each of these missions, the CH-54 could pick up any pod and move it to where it was needed in a few minutes. The CH-54 could then return to its prime mission of moving material to support combat missions. The pods provided more capability with less Cranes.

The Army bought four pods, which were delivered with the first four YCH-54As. *Rotor & Wing* of November 1984 reports, "The CH-54 in Vietnam transported vehicles and heavy hardware over otherwise impenetrable terrain ... one Skycrane lifted a pod with 90 people including 87 combat equipped troops in April 1965 ... no other [American] helicopter has carried as many troops at once."

Pod Missions

The Army CH-54 handbook lists the following missions for pods: troop carrier, communications, first aid station, Red Cross disaster shelter, forward air control van and quarters, aircraft repair and sleeping quarters, mess hall, operations, and mobile ordnance, among others. The first four pods were successfully applied in Vietnam, and the Army ordered 22 more pods to go with the next order for CH-54s. A pod is shown in a combat area in Fig. 5.8.

Fig. 5.8 CH-54A with Army pod in forward combat area (note soldier with weapon handy).

CH-54s SEE ACTION IN VIETNAM

Frank J. Delear, public relations manager at Sikorsky Aircraft Division, reports the following in the May 1966 issue of *Vertiflite* about the four Cranes shipped December 1965:

> It didn't take long for the CH-54s to prove their versatility and attract attention in Southeast Asia ... a New York Times correspondence reported from An Khe, South Vietnam, "Of the 434 aircraft in the First Cavalry division [Airmobile division] these are the favorites.... The Flying Cranes can do astonishing things and in the few weeks they have been here they have impressed everyone."
>
> Among the Skycranes earliest combat area activities were the airlift of enough C-rations to feed a battalion for three days, and hauling a 105-mm howitzer, its crew, and ammunition into battle. Each operation required only hours ....
>
> Shortly after the First Cavalry Division reached Vietnam, a Skycrane carried a bulldozer to the top of a 1200-foot mountain whose slopes were covered with a dense rain forest. The bulldozer was needed to clear the mountain top and position all the equipment necessary to build a radio relay station.... The total was done in one day instead of two weeks ... Major T. J. Clark ... in charge of the 478th heavy lift outfit ... comment[ed] ... "the Army ordered six flying Cranes from Sikorsky Aircraft, which developed the big helicopters in the hope that someone would recognize its utility."

IGOR SIKORSKY PROVIDES A BARGAIN FOR THE ARMY

An analysis of current Army procurement for a new type of helicopter indicates that it can take the Army eight to ten years to study, define, specify, conduct a competition, contract for development, and finally produce the first of a new helicopter. The development time would include the design, construction, and testing of several helicopters from each of two competing companies and thorough flight testing by the companies and Army of both of them, before the Army selected one. The chosen design would then be tooled up to produce the first serviceable one. This process would typically take many hundreds of millions of dollars and has been labeled "fly before you buy."

The "fly before you buy" procurement policy was adopted after the unsuccessful Army procurement of the Lockheed AH-56, which was to meet the Army's armed aerial support system (AAFSS requirement). The award decision for Lockheed was made after the Army evaluated paper proposals from two contractors. This saved time and cost by awarding just one development contract, but after two of the Lockheed gunships were destroyed and a pilot killed in its development effort, the program was cancelled.

It appears that Igor Sikorsky saved the Army eight or more years and many hundred million of dollars by defining and developing the S-64 Skycrane. The first of six Army YCH-54s was built and delivered to the Army under a $13,000,000 fixed-price contract in one year. In another half-year, four were sent by ship to support Army combat operations in Vietnam. As soon as the ship arrived, YCH-54s began moving cargo from ship to shore. Moving cargo this way was much faster than unloading the ship at the dock, loading trucks to transport material inland, and then unloading trucks. Eighteen months and $13,000,000 compare well to eight years and many hundreds of millions of dollars for a helicopter that "could do what no other helicopter could do" according to one Army pilot.

The CH-54s set a record with the amount of engineering equipment that was moved during operations in Vietnam. Some of the equipment moved by the CH-54s were bulldozers, backhoes, and trucks. The CH-54s even moved 155-mm howitzers over 100 times to support active combat operations.

CH-54 MISSIONS

In 1968, the first CH-54s from the production line went to Vietnam. Then Major Gary R. Heffner (retired as Lieutenant Colonel) assumed command of the 273D Assault Support Helicopter Company (Heavy), the first heavy-lift helicopter company in Vietnam with a full complement of six CH-54s plus two maintenance float CH-54s. He developed operating procedures and kept detailed records of the 273D missions, loads, and other important measures, including a typical mission profile pictured in Fig. 5.9.

Almost all missions were in support of combat operations, including those of Army engineers who were often building forward fire bases (FFBs),

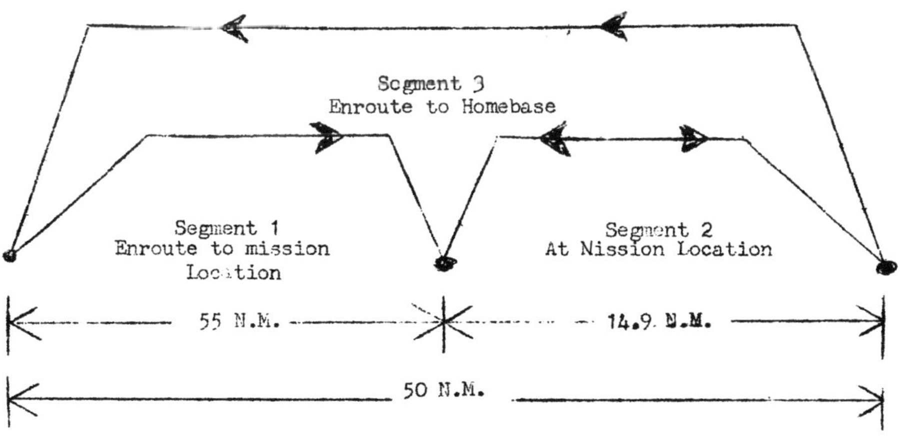

Fig. 5.9 Typical mission profile [courtesy U.S. Army, provided by Lieutenant Colonel (retired) Heffner].

sometimes under fire. An unidentified newspaper article states that "The An Khe Forward Fire Base depended 100% for helicopter support to maintain its operation in hostile enemy territory. The CH-54s were used for the heaviest loads." FFBs were part of a strategy to "take it to the enemy" by locating armed bases in the middle of enemy territory to engage the enemy. FFBs were created by Army engineers, which used CH-54s to fly in bulldozers, artillery, ammunition, fuel, food, sand bags, concrete, observation towers, etc. to build the base. FFBs then depended on support from various helicopter types to survive.

CH-54 OPERATIONAL DATA

Figure 5.10 provides data for 105 days ending 15 June 1968, such as flying hours, sorties, and tonnage. Distribution of tonnage, visible at the bottom of the table, reveals that 70% of the tons were moving ammunition. This was mostly for combat units preparing to or engaging the enemy. The 21% for engineer equipment was primarily for building FFBs and bridge repair, which were combat installations (see Figs. 5.11–5.13). Most of the 4% for artillery was moving 155-mm artillery into difficult terrain to support combat. The 8% for combat vehicles was supporting combat units as was the POL (petroleum, oil, and lubricants).

TABLE II
GENERAL DATA
1 March 1968 to 15 June 1968

	MARCH	APRIL	MAY	1-15 JUNE	ENTIRE PERIOD
Flying Hours - Total	316.6	308.4	291.7	216.2	1132.9
Mission Flight Hours	301.8	292.1	274.3	205.5	1073.7
Test Flight Hours	14.8	16.3	17.4	10.7	59.2
Operationally Ready	36 %	46 %	63 %	63 %	51 %
Not Operationally Ready - Maintenance	19 %	18 %	24 %	21 %	20 %
Not Operationally Ready - Supply	45 %	36 %	13 %	16 %	29 %
Missions - (A Grouping Sorties requiring commitment of at least one Aircraft)	45	47	38	24	154
Average Mission Flight Time - Hours	6.7	6.2	7.2	8.4	7.0
Average Number of Aircraft Committed Daily	1.4	1.5	1.2	1.6	1.4
Sorties	803	855	813	563	3034
Tasks	394	440	390	277	1501
Tonnage Transported - STON	2623	2959	2612	1826	10,020
Ton - Miles Flown	47,756	37,225	35,688	28,529	149,198
Distribution of Tonnage					
Ammunition (Class V)	77 %	68 %	73 %	60 %	70 %
Engineer Equipment	12 %	15 %	10 %	15 %	13 %
Artillery	2 %	3 %	5 %	6 %	4 %
Combat Vehicles	5 %	10 %	5 %	13 %	8 %
POL	1 %	1 %	1 %	----	1 %
Aircraft Recoveries	1 %	1 %	2 %	----	1 %
Miscellaneous	2 %	2 %	4 %	6 %	3 %

Fig. 5.10 Actual mission data [courtesy U.S. Army, provided by Lieutenant Colonel (retired) Heffner].

Fig. 5.11 Lifting base of observation tower to deliver to an FFB [courtesy of Lieutenant Colonel (retired) Heffner].

One measure of productivity is the operationally ready rate expressed in percentage in line four of Table II (Fig. 5.10). It increases sharply from 36 to 63% in 3½ months. One of the major reasons for this was that the not operationally ready-supply entry decreases from 45 to 16%—by more than half. Lieutenant Colonel (retired) Heffner explained that this occurred because the aircraft arrived and became operational before the full supply of spare parts was delivered.

A comparison of tons moved per day between commercial and Army operation shows higher productivity for commercial operations, but they operate in a different environment for the Army. One commercial Skycrane flew

Fig. 5.12 Placing a water tower on its base [courtesy of Lieutenant Colonel (retired) Heffner].

Fig. 5.13 Precisely placing a new bridge section to replace a section blown by the enemy.

300 hours in 30 days, averaging 10 hours per day, in a logging operation, using two flight crews in highly planned and controlled conditions. The log cutting started weeks ahead, and the logs were waiting for pickup. The total cycle time for 8000 pounds of logs was usually four flight minutes or less. The fuel truck was also just a few flight minutes from the logging site so that frequent and rapid hot refuelings were quickly executed. A maintenance trailer was also nearby, so maintenance and repairs could be done at night.

The Army Skycrane was on-call for combat support missions, which were unpredictable, regarding the day, time, load, pickup point, delivery point, and refueling location. On some days, adverse weather precluded flight and mission accomplishments. Evasive flight paths were often required because of enemy action. Even so, CH-54s and their crews took many hits from enemy fire. Productivity is normally a measure of predictable and repetitive actions, which the Army does not see in combat situations. No wonder commercial productivity is higher than the military experience. They operate in different worlds.

CH-54A Lifts Over 42,000 Pounds

The mission analysis shown in Fig. 5.14 (in Heffner's handwriting) covers 443 flights during a 3½-month period in 1968, averaging about 31 lifts per week, that were flown over the approved maximum gross weight of 42,000 pounds for the CH-54A. Heffner explains,

> These higher gross weight flights were made during the exigencies of combat—missions where you either got the job done or comrades were

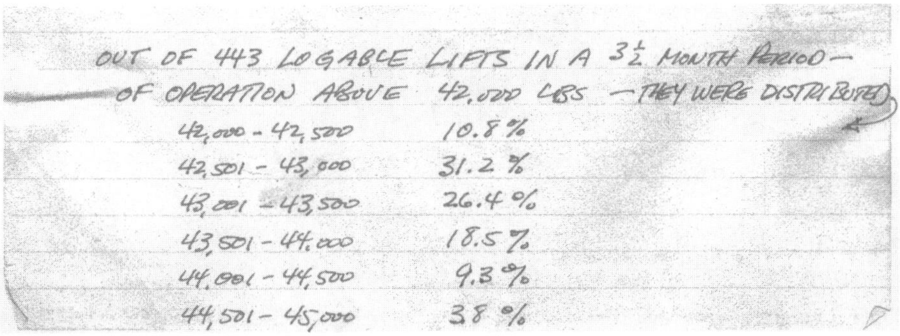

OUT OF 443 LOGABLE LIFTS IN A 3½ MONTH PERIOD—OF OPERATION ABOVE 42,000 LBS —THEY WERE DISTRIBUTED

42,000 - 42,500	10.8 %
42,501 - 43,000	31.2 %
43,001 - 43,500	26.4 %
43,501 - 44,000	18.5 %
44,001 - 44,500	9.3 %
44,501 - 45,000	3.8 %

Fig. 5.14 Lifts over 42,000-pound gross weight limit of the CH-54A in one 3½-month period [courtesy of Lieutenant Colonel (Retired) Heffner].

> put at higher risk—you did what the situation required, as safely as you could—period. Obviously flights under these conditions took extra care and attention. One had to be ahead of the aircraft, more power was required, and smoother control technique was necessary. The Flying Cranes handled the demands well. Extra and more stringent inspections were made of the airframe, rotors, and engines, but no observable damage resulted and no long-term damage became evident.

Lieutenant Colonel (retired) Silva, who was commander of the 273D after Lieutenant Colonel (retired) Heffner, states that the Final Report of Operations of the 273D Assault Support Helicopter Company (Heavy) 28 October to 1 February 1969 stated that over 95% of assigned missions were accomplished using the single-point suspension system and ammunition resupply represented 58% of total tonnage handled.

RECOVERY OF DOWNED AIRCRAFT

Figure 5.10 lists aircraft recovery as a mission taking a minor amount of flight time, but it got quite a bit of newspaper coverage. It also saved many hundreds of millions of dollars of aircraft, which would not otherwise be recovered. Many of these aircraft were also returned to service. An article in another unidentified publication cites a study that states the following:

> Flying Crane helicopters ... recovered a total of 357 allied aircraft downed in Vietnam, valued at almost 200 million dollars.... The total value of the aircraft recovered as of June 1967 was about 100 times the cost of a single CH-54A. One hundred of the recovered aircraft were valued at approximately $120 million and could not have been retrieved by any other Army helicopter. Most of these aircraft would have been lost or destroyed. Sixty two of these aircraft were Army CH-47s, the largest Army troop transport helicopter.

a) b) c)

Fig. 5.15 Three recoveries: a) retrieval of Air Force fighter/bomber, b) retrieval of Navy fighter/bomber, and c) retrieval of Army CH-47.

A later report of recoveries in 31 December 1969 listed 587 aircraft recovered with a value of \$342,090,000. The Vietnam War ended about three years after the year of this report, and so the total number could have been 1047 aircraft retrieved, if it continued at this rate. This could have amounted to over \$610 million retrieved during the entire war, if the value of the aircraft retrieved continued at the same rate (\$342,090,000 divided by 587 aircraft times 1047 aircraft equals over \$610,000). Three recoveries are shown in Fig. 5.15.

Another unidentified report reveals an active week for the CH-54s: The cranes performed a large variety of operations, including recovering six UH-1Ds, a CH-47, and a bulldozer; transporting containers of fuel; and transferring the medical pod and the command post to another location.

RETRIEVAL OF A DOWNED SKYCRANE

A 1970 article from *HAWK* magazine, "Skycrane Down!" about the "Super Hooks" of the 273D, illustrates how challenging a single retrieval was. One of the company's CH-54 Skycranes was forced to land 160 miles south of Long Binh. Two hundred and sixty soldiers were immediately flown in to secure the ship, overnight, while recovery plans were being formulated.

Early the next morning, a team of maintenance personnel were brought to the site. They are shown in Fig. 5.16 removing, with their muscle power, one of the two EAPS, which weighed about 300 pounds, located in front of one of the engines. They also prepared the rotor head for removal from the downed Crane by the rescue Skycrane, which picked it up and gently placed it on the ground. The blades were removed from the rotor head and were placed with the rotor head in another helicopter to be returned to base. The rescue CH-54A then hovered over the stricken ship and lifted it from the mud to fly it back to its base (Fig. 5.17). The injured helicopter had to be landed four times so that the rescue helicopter could refuel in order to make the 160 miles with this heavy load.

Fig. 5.16 Manpower removes one of the 300-pound EAPS [*HAWK* Magazine, courtesy U.S. Army, provided by Lieutenant Colonel (Retired) Silva].

Fig. 5.17 The downed Crane is lifted and flown to its home base [*HAWK* Magazine, courtesy U.S. Army, provided by Lieutenant Colonel (Retired) Silva].

CH-54 LOAD CALIBRATION REDUCES ACCIDENTS

Lieutenant Colonel (retired) Silva reported for the book that several CH-47s had accidents while lifting loads on an external sling because the weight of the load had been underestimated. Each CH-54 had a special load-measuring device, called a tensionometer, which was calibrated each morning against a 10,000-pound concrete block at their base. To reduce the accidents, a Skycrane would fly to the CH-47 loading area each morning and lift, for a few minutes, the heavy loads that the CH-47s were to transport. This identified the loads that were within the CH-47 capability, and the number of CH-47 accidents was reduced.

TRAINING COMBAT FORCES IN THE USE OF THE CRANE

Initially, the combat units were not fully trained to effectively use the CH-54s in estimating the correct weight of the load, learning safe hookup methods, planning of multiple CH-54 missions, using the range of the aircraft, etc. Heffner issued a memo on 1 March 1968 to familiarize operating units with CH-54 airspeed, endurance, and lift capabilities for transporting ammunition, fuel, artillery batteries, engineering equipment, aircraft retrieval, and even retrieval of cargo from ships.

The different lift modes, such as single-point, four-point suspension, and 32-point hard point capability, were explained as was use of the pod. Load weights and ranges of flight also were outlined. Management of the static electric charge from grabbing the hook, which could knock a man unconscious, was explained. The CH-54 had no armament, and so gunship support was needed where enemy fire was anticipated. Heffner provided a checklist (Fig. 5.18) to cover on-site conditions.

OVERLOAD OF THE CH-54A CAN CAUSE ACCIDENTS AND LOSS OF LIFE

Despite this effective planning, Crane pilots still had to be alert for dangerous surprises when lifting any load. Heffner recalls one routine lift of timber that could have cost lives:

> The task was to transport a four-point load of lumber. The load looked heavy, so I de-planed to talk to the senior Sergeant who assured me that the load had been calculated and was not greater than 15,000 pounds. He said he had checked the manuals himself to confirm the weight.
>
> We positioned over the load and made a four-point hookup. When I took off, the load seemed heavy and the initial climb took a bit more power than normal, but not so much that it alarmed me. By this time, I was just through translation lift and in seconds was into blade stall. I couldn't stop the climb-out and couldn't jettison the load, because it was not on the hook. The only place to land was behind the aircraft. I slowly did a shallow left turn to return to the landing area.

DEPARTMENT OF THE ARMY
273D ASSAULT SUPPORT HELICOPTER COMPANY (HVY)
APO San Francisco 96291

1 March 1968

CHECKLIST TO INSURE MAXIMUM USE OF CH-54 FLYING CRANE

1. Pick-up and drop-off zones, well policed and located away from tents and situated so that the CH-54 can approach and depart into the wind without over flying the camp.

2. Serviceable rigging material must be used to preclude failure in flight and subsequent loss of load.

3. Hook-up personnel must be equipped with protective mask or goggles and leather work gloves. They must be well briefed to eliminate their fear of the cargo hook thus facilitating the cargo hook-up.

4. Sufficient prepared loads of 14,000-16,000 lbs to keep the CH-54 working without delay. If there are lighter loads to be carried program them to be carried first when the CH-54's fuel load is greatest then move the heavier loads on subsequent trips as the CH-54 burns off fuel.

5. All loads must be attached to the CH-54's cargo hook with a metal clevis.

6. All items carried on equipment being moved (i.e. equipment in the bed of 2½ ton trucks, equipment on trails of 155 howitzer) must be properly secured so that it arrives at its intended destination.

7. Call-signs and frequencies must be the same ones listed on the mission request. Personnel should be on the radio 30 minutes prior to scheduled reporting time. Alternate call signs and frequencies if available should be listed on the mission request.

8. Smoke grenades should be available at the P/U and D/O zones.

9. Single point pressure refueling nozzle (FSN USAF 4730-289-0096) should be provided in refueling areas to speed-up refueling operations and reduce maintenance problems.

10. Try to overwork the CH-54 and its crew - WE'LL BET YOU CAN'T DO IT.

Fig. 5.18 Checklist used to ensure maximum use of the CH-54 [courtesy U.S. Army, provided by Lieutenant Colonel (retired) Heffner].

A right turn would have taken less power, but I was in the left seat and I could see better to land from that seat. At 150 to 200 feet high, I continued a shallow bank and completed a race track orbit to position the aircraft for landing. I was in virtually constant blade-stall, but was able to complete the turn and get the aircraft on the ground successfully, because I could make a roll on landing. If the lift had been from a small clearing, we would have lost the Crane and perhaps some lives.

We unhooked the heavy load, checked the aircraft for damage and found none. I then found the Sergeant to find out how the mistake had been made. The Sergeant showed me that the manual, which referred to regular pine and similar construction timber. Our load consisted of tropical hardwood, which was much more dense and heavier than pine. We split the load up and moved it in two lifts. I did not make that mistake again.

> The CH-54 was great to fly, because it gave you a feeling of accomplishment, being able to help the combat units in a way that no other helo could. I never had a bad moment due to the Crane's performance. There were a few situations, if it had not been for God and the soundness of the Crane, I would not have survived.

SPECIAL MISSIONS

Because of the unique capabilities of the CH-54, it was often called on to perform unusual missions compared to its regular duties of moving of artillery, ammo, fuel, trucks, bridges, medical pods, etc. These missions were ones that no other helicopter could perform. One mission was to lift large objects from a ship at anchor, or from a ship underway. Another mission required removing a damaged precision part from an artillery piece, flying it to the repair site, and returning it for precise reassembly, while hovering over the artillery piece in the combat area. In another case, the CH-54 was called to lift an armored personnel carrier (APC) off a soldier who had been pinned under the APC after the explosion of an enemy mine.

Completing these missions required the unique features of the CH-54, such as the aft-facing pilot, who had excellent visibility of the target and could control the aircraft (up and down, left and right as well as forward and backward). He also controlled the hoist and hook to complete the mission. The hoist and hook could reach 100 feet below the aircraft into a deep jungle or into the hold of a ship to deliver or retrieve critical objects of any size within the weightlifting capacity of the CH-54. Some of these special missions and related events follow.

SHIP-TO-SHORE OFFLOADING OF AUSTRALIAN CARRIER BY A CH-54A

Another article describes the use of a CH-54A to offload an Australian aircraft carrier anchored offshore in Vietnam because no dock was available. Sikorsky Aircraft had previously tested this concept using its commercial version of the CH-54A to move 30 full-sized containers from a container ship five miles offshore from a Sikorsky facility in Bridgeport, Connecticut, to prove the speed of the "ship-to-shore" concept.

More than 224 tons were moved from the Australian aircraft carrier HMAS Sydney off Vietnam by one CH-54A, which delivered a load every nine minutes. Lieutenant Colonel (retired) Heffner, who piloted these flights, explains that the plan began with Lieutenant Colonel Ian Gilmore of Australia, who wanted to accelerate the delivery of material to his troops at Vung Tau (Fig. 5.19):

> We worked together to develop a plan to move the material using one CH-54A. The loads were containers, vehicles and general cargo. The loads were to be moved about two miles from the ship directly to the

Fig. 5.19 CH-54A offloading Australian aircraft carrier off Vung Tau, Vietnam (*Blackjack Flier*, Vol. 1, No. 5, Feb. 1968; courtesy U.S. Army).

> Aussie base. On many of the return legs the Crane moved cargo to be returned to Australia. This proved the feasibility of rapid ship to shore movement of heavy military cargo.

COMMERCIAL PROPOSAL FOR SHIP TO SHORE FROM UNITED STATES TO VIETNAM

With ships heavily backlogged in Vietnam and critically needed military equipment delayed, the Isbrandtsen Shipping Company submitted a proposal to the Army for a turnkey system including seven ships and seven onboard Skycranes. Isbrandtsen forecast the Skycranes would reduce by 40% the total turnaround time for each shipment. The time savings came from two reduced loading times, one in the United States and mostly in Vietnam straight to Army bases where the material was needed. The commercial Skycranes and ships could be made available in reasonable time and had proved the concept previously. The proposal also forecast a saving of millions of dollars for each roundtrip of each ship. It was not approved.

SHIP-TO-SHORE MOVEMENT OF CRITICAL LOADS FROM QUI NHON HARBOR

Another unidentified article reports that fifteen Army Conex containers were taken inland by a CH-54 making eight trips from an aircraft carrier in Qui Nhon harbor. Some of the CH-54s loads were extremely heavy, like 17,500-pound bulldozers, 17,000-pound Conex containers, and a 17,000-pound signal van.

EXTRACTION OF HIGH-PRIORITY SINGLE LOAD FROM CONTAINER SHIP

Moving these heavy loads from the container ship to the shore was not always an easy task. The CH-54 pilots sometimes had to use intricate maneuvering to hook the cargo before even thinking about moving it to a location onshore. Lieutenant Colonel (retired) Heffner explains the following for one such maneuver:

> With all dock space in the Mekong River occupied to Saigon, Headquarters called for a Crane to move a high priority cargo from the

ship-to-shore location. The acceptance of the assignment was conditional, since the mission was not well defined. We then flew out to the ship and established communications. The container was 12,500 pounds, but in a difficult location on the ship. This meant it had to be extracted vertically and then horizontally, so the tail rotor would not hit the mast of the ship.

I had to hover with about a 75 foot of cable to make the hookup and had difficulty positioning the hook. I then realized I was positioning the Crane by using the mast of the ship, which was rolling with the waves. When I corrected my hover reference, the hook stabilized and the crew was able to make the hookup. I raised the load and flew it to its destination.

SHORE-TO-SHIP MOVEMENT OF DAMAGED AIRCRAFT

Not only did CH-54s carry equipment from ship to shore, but another unidentified article states that CH-54As transported damaged Marine UH-34/S-58 helicopters to Navy vessels waiting offshore. This raises the possibility for the Navy to use CH-54As to move damaged Navy aircraft from an aircraft carrier onto Navy ships returning to the states. This would also clear badly needed space on the carrier, so that replacement aircraft could be flown onto the carrier to bring it back up to full strength.

Fig. 5.20 CH-54A removing cargo from LSM while underway [courtesy of Lieutenant Colonel (retired) Heffner, who was the pilot].

Ship-to-Shore Movement from Ship Underway

In addition to moving equipment from ship to shore and back again, CH-54s also moved equipment from ships already underway. Working with the Australians, Lieutenant Colonel (retired) Heffner reports, "With the success of the prior ship off loading, we worked with the Australians to test off loading a ship while underway. This had never been done before to our knowledge." Using an Australian lighterage vessel, LSM, the load would be removed while the ship was underway to prove extraction capability, as shown in Fig. 5.20.

Again, Heffner reports, "This required the LSM to head up wind at 15 to 20 knots as the Crane operated above the stern of the ship. Relatively constant winds and a low gust spread were needed. The test was completed successfully and proved the CH-54 could remove cargo from a moving ship deck under the right circumstances."*

Moving 175-mm Gun Tubes for Repair

The CH-54As were called upon to perform another task: moving gun tubes for repair. Lieutenant Colonel (retired) Heffner explains that

> The 175-mm artillery piece could fire a shell a long distance, but its accuracy declined rapidly as the tube wore. Reboring the gun tube on a large reboring machine can restore accuracy, but the machine was a long way from the combat areas and the normal gun positions. CH-54s were called upon to pick up the gun tubes to deliver them to the repair facility and to return them and position them on the gun. After a suitable sling was rigged, the tubes were delivered to the facility and back to the guns successfully.

Single-Engine Flight from Combat Zone

The CH-54A was also able to fly successfully on only one engine. This ability was tested on a resupply mission to an outpost within a few miles of the Ho Chi Minh Trail, near Cambodia, by Lieutenant Colonel (retired) Heffner, who reports the following:

> We were unable to start one engine for the trip home. For security reasons, we could not remain in this area for the time it would take for a repair team to come from our home base. We had the fuel to get home on one engine, but I would have to fly at an altitude that would allow a successful autorotation landing to a secure area, if we lost the remaining engine. We used the pilot handbook to find the altitude and planned a

---

*The details of this test are to be found in Major Heffner's five-page memo and associated photographs of 26 March 1968.

> route to be near secure areas on the way home, in the event the last engine stopped and we had to perform an autorotation. The altitude was calculated to be 13,500 to 14,000 feet. I took off with the single engine and climbed to about 5000 feet. I then continued the single engine climb en route to 13,500 feet and returned successfully to home base.

NIGHT-INSTRUMENT FLIGHT

Nighttime was a challenge for CH-54A pilots, but not one they could not overcome. The ability to fly solely on instruments was another benefit of using the CH-54A. Lieutenant Colonel (retired) Heffner describes a night-instrument flight as follows:

> While on a mission south of Danang, we ran out of daylight. We needed to get to a secure base because our current location would be dangerous at night, when the local farmers in that area turned into Viet Cong fighters during the night. I had previously used the air traffic controllers at An Khe and had confidence in them. Under the deteriorating conditions, this was important, so we took off and headed to An Khe about 50 miles away.
>
> The overcast prevented any reference to stars and there were no ground lights in this remote area. We flew on instruments all the way to An Khe. It was a demanding fight in the clouds and rain most of the way. The AFCS [automatic flight control system] functioned well and was a definite asset on the flight. I was able to establish radio contact with air traffic control and completed a GCA [ground-controlled approach] to the runway and hovered over to the 478th revetment.

NOT EVERY MISSION WAS COMPLETED

Although the Crane had many successful missions, occasionally something malfunctioned, and the pilots had to improvise. Lieutenant Colonel (retired) Eldridge W. Brock, CH-54A pilot and former commander of the 478th Aviation Company of CH-54As, explains that when he was taking off from a remote base and refueling station, he

> . . . lost power and had to auto-rotate into a nearby dry rice paddy. We got some ground troops to surround and protect the Crane over night and then we learned that the refueling station had gotten water in the fuel.
>
> The next morning we got a bulldozer to smooth out a towing path back to the remote base. We towed the Crane with an armored personnel carrier using nylon straps connected to the Crane. A portion of the trip was down a slope, so we restricted the Crane movement with second personnel carrier and straps. The arrangement worked beautifully. On the third day, we brought in a maintenance officer and maintenance crew from our home base at An Khe. They drained and flushed out the contaminated fuel and flew the Crane back to An Khe, where it was returned to normal operation, after a thorough inspection.

What he did not mention was that he was awarded the Bronze Star Medal. The citation stated that he "exposed himself to heavy enemy fire as he quickly established a defensive perimeter around the disabled aircraft... remained with the helicopter throughout the night, supervising repairs and proceeding to make plans to evacuate the damaged aircraft to safety...."

Not Every Crane Returned

Not all pilots were able to recover the aircraft after a major malfunction. Lieutenant Colonel (retired) Brock reports on an accident that occurred after smoke was seen coming from the Crane as follows:

> A pilot was in a hover and preparing to enter forward flight, when ground personnel saw smoke and signaled the pilot to land. The pilot landed, cut one engine, and applied the rotor brake to stop the rotor. He failed to lock the nose wheel, and the torque induced by stopping the rotor caused the Crane to turn and roll down the nearby slope. The wheel brakes did not respond and Crane headed to a nearby aircraft, picking up speed. It missed the aircraft, but continued down the slope and crashed in a small ravine. The pilot survived with a broken leg and other injuries.

10,000-Pound Helicopter Bomber

In addition to its duties as a heavy equipment mover, the CH-54 temporarily became a bomber, but that role did not suit the CH-54's design. Chief Warrant Officer (CWO 4; retired) James R. Oden explains that

> I was preparing for my second tour to RVN [Vietnam], when I received a call for a mission that could not be discussed on the phone. The objective was to create an instant LZ [landing zone] where troops and artillery could be placed to combat the enemy based in triple canopy trees that hid and protected them. The commanders wanted these LZs on the highest terrain available for maximum advantage. The plan was to adapt the CH-54 to carry a 10,000 bomb that was WWII surplus. If successful, a team of Army Engineers would immediately land in the LZ and clear the area for other helicopters to bring in the troops.
>
> At Kirkland Air Force Base, a cradle was fabricated to attach the bomb to the CH-54. A special harness lifted the bomb tightly against the cradle, a three-foot pipe contained the control that triggered the arming mechanism at the nose of the bomb. A WWII drift meter was added to the aft cockpit, so the drift of the aircraft could be corrected. The aft pilot became the bombardier and would pass drift corrections to the pilot. On the first inert bomb test at Kirkland, three smoke passes were made before the inert bomb was successfully dropped.
>
> The first armed drop was in RVN. The release was successful and we pulled full power to avoid being hit by shrapnel from the bomb. Other

> bombs were dropped in other areas, but only one was useful. I flew into that LZ a few days later to retrieve a downed helicopter. The bomb had done a good job of clearing the LZ. The Air Force continued to drop 10,000 pound bombs from the back of C-130s. I am not sure of the degree of success. As for the CH-54 bomber, it was returned to standard operational configuration.

Movement of Navy "Swift Boats"

The Navy needed to move a dozen river patrol boats, called Swift Boats, from one river to another. This meant a long cruise by the boats down to the sea, a trip in the ocean to the mouth of the next river, and then a long trip up that river to the destination. Lieutenant Colonel (retired) Silva states that

> Approval was obtained to use a Crane to move the boats. Once the slings were arranged, the trip became routine, except for one of the last trips. The Crane crew member in the aft pilot seat noticed that they also had two passengers, who were waiving at him from the boat as they flew to the next river. The sailors had hidden in the boat before liftoff and were now sitting in the boat enjoying the ride.

Flying Truck

Even though the Cranes fulfilled serious and necessary missions, the next anecdote reveals one pilot's "lighter side." Lieutenant Colonel (retired) Silva relates a comical story about one of the less exciting transport missions:

> When a Crane was flying with a truck as a suspended load during a routine mission, its pilot heard the radio call of a USAF C-130 approaching his area. Hoping to add a bit of fun during a dull flight, the Crane pilot climbed his Crane into a nearby cloud with the truck suspended just below the cloud and continued his flight. He was pleased to hear the shocked Air Force pilot report to his base that he could see something very strange—"a truck flying near my aircraft."

Crane Lifted an APC Off a Trooper

The CH-54 was also used to help injured soldiers when equipment was too heavy to move without the strength of a Crane. Lieutentant Colonel (retired) Silva explains an incident where the power of the CH-54 was put to good use helping an infantryman:

> An Armored Personnel Carrier had struck a land mine and had been flipped over onto a soldier who had been riding on top of the APC. The Crane's rear facing pilot controls were used to lift one corner of the APC in order to successfully extract the injured infantryman.

Crew Exposure to Combat Conditons

Even though the Cranes were not part of combat, the Cranes were often exposed to ground fire. Lieutenant Commander (retired) Heffner relates the following story:

> The Cranes were frequently hit by small arms fire and often returned to base with holes in the fuselage and the blades. On one flight, the crew chief was hit in the forearm. One pilot of the 478th CH-54 Company lost a leg. On one of my flights, several rounds went through the cockpit about a foot from my head.

But Cranes also came under fire while on the ground in forward areas. An article (from an unidentified newspaper) entitled "Major Saved Flying Crane from Attack" is summarized here. Major Gary Heffner, pilot of a CH-54 in Vietnam, got into a trench when mortar fire started falling on his airfield in the demilitarized zone and then led his crew back to start his helicopter in a record two-minute time as shells were landing close by. He received an Army Commendation Medal with V for his quick actions and bravery.

Operating Room Lifted and Flown into Battle

The 25 February 1966 issue of *LIFE* magazine had an article about a medical operating room—a medical pod—being lifted and flown into battle by the CH-54[†]:

> At its home base at An Khe, a "flying crane" helicopter swoops down in a swirl of dust. Clutching onto a mobile surgical "pod" it soars away on an urgent medical mission into the thick of battle.
>
> The fatality rate among the wounded in Vietnam is far below that of World War II and Korea, largely because the techniques of evacuation of casualties have been keyed to the new concept of mobility. The pod is an emergency clinic, with X-ray, laboratory, and all-purpose surgical facilities. Within minutes of being hit, a man can get surgical care even before he is evacuated.

Figure 5.21a shows a large new Sikorsky pod flown to Fort Benning for Army flight tests in 1963. The pod is much wider than the stick fuselage that carries it. The stick fuselage is minimized to carry external loads, but the pod is very wide to carry very large internal loads. It can seat 87 fully equipped soldiers or accommodate a full medical operating room. It had the largest "fuselage" of any helicopter in Vietnam or of any U.S. helicopter built to date of this writing. The successful tests of the first pod led to delivery of the first of twenty production pods as shown in Fig. 5.21b.

[†]Article was provided by Jay Rickmeyer.

Fig. 5.21a Large Sikorsky pod is delivered to Fort Benning in 1963 for Army test.

Fig. 5.21b First Army pod from Sikorsky ready for delivery.

CH-54A WORLD RECORDS

In his article "The Crane Is Almost Extinct" in the December 31 1996 *Army Aviation* magazine, Lieutenant Colonel (retired) Paul J. Fardink, a former CH-54 pilot and writer of many aviation articles, writes,

> On 30 December 1968, after several different attempts, using Sikorsky's published best rate of climb charts, Chief Warrant Officer James P. Ervin remarked "I thought it over and decided that the cleanest air you can get is that coming through the top of the [rotor] disc. So I decided to make a totally vertical ascension." Ervin piloted the ship straight up at vertical

> speeds at over 6000 feet per minute, switching to Sikorsky-published climb data at 20,000 feet. When Ervin went through 30,000 feet, he was still climbing at 995 feet per minute, but it was too late and dark to try further.

Ervin set three new world records: 1 minute 36.8 seconds to 3000 meters, 3 minutes 31.5 seconds to 6000 meters, and 7 minutes 36.4 seconds to 9000 meters. On an earlier flight that day, he had established the horizontal flight-at-altitude record of 31,280 feet. Ervin states, "Air Traffic Control radioed a commercial airliner flying at 17,000 feet to be advised there is a helicopter at your 9 o'clock position descending out of 27,000 feet at 4000 feet per minute." Ervin said there was a brief pause where the airline pilot exclaimed, "Good Lord, you mean they are up here right now?" Another pilot was heard to ask, "What kind of helicopter is that?" Fardink adds, "Only the U.S. Navy F-4 has a faster time to 3000 meters at this time." The more powerful CH-54B set higher records at a later date.

CH-53 Recovery in Vietnam

In the preceding article, "The Crane Is Almost Extinct," Lieutenant Colonel (retired) Paul Fardink also explains that a CH-54 recovered a downed Sikorsky/Marine S-65/CH-53, the largest helicopter in Marine service and very close to the size of the Crane itself. There was concern by the Army Crane pilot Edward M. Strazzini that it might be too heavy:

> When we arrived, the Marines assured us that they had removed the rotor system, transmission, and some other minor components to get the weight down to 17,000 pounds ... when we made the initial pickup and hover check, we were surprised to see our winch-load indicator telling us that we had in excess of 20,000 pounds ... the daylight was fading.... So we held a hover for about 30 minutes to burn off another 2000 pounds of fuel, giving us the absolute minimum in our tanks needed to get us to the Marine base with the load ... after landing, I walked over to examine the '53 and was shocked to discover that the recovery crew had loaded much of what they had removed from the aircraft into the cargo bay of the CH-53A!

Conclusion

CH-54 Performance and Maintainability

Major Heffner provided valuable data in his 24 June 1968 "Report and Analysis of CH-54 Heavy-Lift Helicopter Operations and Maintenance." Paragraph b.4 Maintainability states the following:

> After almost 3 years of use in RVN, the CH-54 has established the reputation of a rugged, powerful, and reliable helicopter. Although

maintenance problems are encountered, none have caused injury to personnel or damage to the aircraft. In cases where the aircraft sustained battle damage, fast temporary repairs have been adequate to permit its safe return to base for permanent-type repairs.

EAPS Solves Engine Power Loss and Maintenance Problem

After the CH-54As were deployed a short time in Vietnam, the engines showed a power loss, which was found to be caused by the ingestion of sand as the Cranes spent a high percentage of time hovering over the sandy ground. As the sand was sucked into the engine, it caused damage and power loss. Some engines had to be replaced in less than 300 hours of service. Sikorsky Aircraft quickly developed an engine air article separator, named EAPS, that could be mounted in front of the engine to filter out the sand. This fix solved the problem with only a small loss of power. As shown in Fig. 5.22, EAPS was a box about 4 feet long consisting of many twisted plastic tubes that caused the sand to spin away from the front of the engine. This solved a major problem of engine wear and power loss in Vietnam.

Army Concept Team Assessment of the 273D

Lieutenant Colonel (retired) Silva explained that the Army Concept Team evaluated the deployment of the 273D for the period of February 1968 to June 1969. The purpose of the study was to document the capabilities and limitations of the company, as the company successfully supported the elements of the four divisions in its area.

The report was favorable and made the recommendation that "The unique ability of the Heavy Lift Helicopter to perform ship-to-shore cargo movement directly to base should be explored." The report also stated that the highest percentage of sorties were at a distance of 18 miles, 7% were up to 5 miles, and 8% were up to 50 miles. The percentage of the missions that required two or more flight hours and that were successfully accomplished was 95.2%.

Fig. 5.22 EAPS (center) is installed in front of the engine (left); long, low assembly at the EAPS bottom collects the sand from the bottom of the EAPS and blows it overboard on the left (courtesy of Raymond L. Robb).

FUTURE HLH

In Heffner's letter of transmittal of the 24 June 1968 Report and Analysis of CH-54 Heavy-Lift Helicopter Operations and Maintenance, Heffner states, "the report is designed to document the experience of a CH-54 helicopter company so as to favorably influence the characteristics of future Heavy Lift Helicopters [HLH]." Paragraph K2C of his report states,

> Full potential of the HLH [CH-54] has not been realized. Mission employment in troop transport [upon receipt of external pods], bridge emplacement, movement of palletized cargo, and ship-to-shore transshipment will provide a capability, the full use of which, will task the imagination of the most astute operations and logistic planners.

Since Vietnam, the Army has paid for design studies from several helicopter companies for a future HLH that would lift 25 tons as compared to the 12.5 tons of the CH-54B. Lieutenant Colonel (retired) Heffner was asked for his thoughts on some of the important requirements for that concept. He said,

> It seems to me that a future HLH should be capable of lifting and transporting the heaviest piece of battlefield equipment that is considered by tactical Army commanders to be that which will positively influence the outcome of the battle. In other words, if the character of the fluid and fast moving battlefield is such that the location of the decisive battles suddenly shift and cannot be adequately countered by the commander's present distribution of forces, then he must have available the capability to rapidly relocate his decisive elements, whatever they are.
>
> Armored and artillery equipment must be within the lift capability of the commander's helicopter resources, which can transfer that equipment quickly and over hostile and unsecure terrain to the location necessary to bring the battle to a successful conclusion. If the battle isn't won, the rest becomes more difficult. Also, I feel if one has the battlefield capability, the logistical requirement can be accommodated.

CH-54S ARE RETIRED

"The Skycrane Is Retired" article by Ken Kula for *Atlantic Flyer* magazine about 1995 (the year the CH-54s were retired) states the following:

> After more than 30 years of service to army aviation, the Sikorsky CH-54 Skycrane has been retired from military service.... The first type was the CH-54A which was first delivered in 1964. Five years later the more powerful and improved CH-54B became available ... They moved loads in Vietnam that no other helicopter could move to support combat operations and logistics needs. Civilian operators find that the Skycranes are prized for their usefulness in logging, construction, and fire fighting.

Fig. 5.23 Retired CH-54 at Army Museum, Fort Rucker (courtesy of Raymond L. Robb).

See Fig. 5.23 for a photograph of a retired CH-54. There are many still operating in commercial markets worldwide at this writing.

The next chapter explains how the commercial markets were defined by the Sikorsky sales team, which took the S-64 to customers' sites on two continents to show customers how to save money, time, and the environment.

A QUOTE FROM IGOR SIKORSKY

It seems clear that Mr. Sikorsky recognized that the success of his radical helicopter in a dangerous and difficult war was because of the rapid actions and leadership of many levels of Army personnel. One of his quotes supports this: "We the designers and builders of airplanes would be building something useless and worthless, if it wouldn't be for the skill and courage of our airmen." Great credit is due the Army "airmen" such as mechanics, flight engineers, pilots, and senior officers, among others, who learned how to employ this unusual aircraft and improve it in record time, both before shipment to Vietnam and during combat operations.

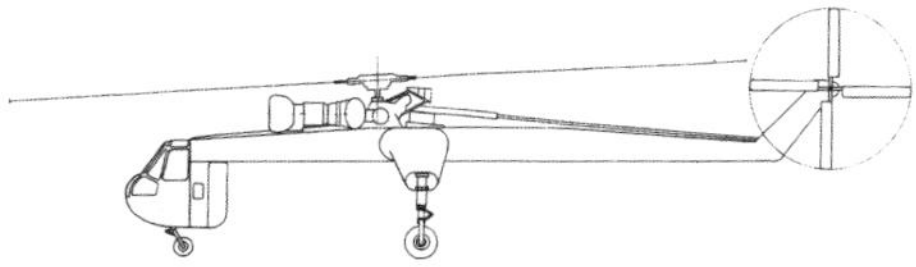

Chapter 6

Sikorsky Develops Commercial Markets

Introduction

As the end of Army CH-54 production could be forecast, E. E. "Tug" Gustafson, one of Sikorsky's senior sales managers, pleaded with Sikorsky management to give him a Skycrane and a team of pilots, mechanics, and other support personnel to prove Skycrane value to commercial customers and sell those already built. Similar demonstrations had been successfully accomplished with the S-60 and S-64 Cranes for military customers and succeeded in obtaining Army orders.

Tug was sure he could sell Skycranes, if he could fly one of the aircraft to customer sites and show customers how to make a profit with the Skycrane. He had expert helicopter knowledge and long experience in the commercial helicopter markets. He finally convinced management to give him a Skycrane and a team of key Sikorsky people to go on the road to sell them. It took four years to sell the first one, and that was not for the forecasted use.

Commercial Crane Sales Team

This team consisted of Tug as director of industrial marketing; Mal Burgess, his operations manager; three pilots; two mechanics; one electrician; and an engineer, as required. Tug was convinced that some of the customers that he sought would see the benefits of the Skycrane and purchase them. Mal would thoroughly plan each different flight operation with the pilots and customers to insure success. The other team members would support the Skycrane to meet the plan. The team believed they would open the commercial markets for the Skycrane.

Flight Demonstrations

Mal Burgess describes the flight demonstration effort: "We took the Skycrane wherever there was opportunity to prove its value. That was anywhere in North or South America. It took several years to demonstrate the different uses of the Skycrane and sell them." The following sections describe some of the key demonstrations.

RAPID MOVEMENT OF RESCUE EQUIPMENT

In one of the first nonmilitary demonstrations, in December 1966, the Skycrane flew a rescue vehicle from Staten Island across New York Harbor to Manhattan to show how quickly scarce rescue equipment could be moved in an emergency across difficult terrain. The views shown in Fig. 6.1 were taken from the aft-facing pilot's cockpit. They illustrate the excellent visibility of the load (see the cross on its top) and also of the hook from this cockpit.

SHIP TO SHORE FROM CONTAINER SHIP

In January 1967, one Crane flew 30 cargo containers, some as heavy as 10 tons, from the American Export Isbrandtsen Lines container ship, just offshore Bridgeport, Connecticut. Seas were running eight feet high, and winds gusted up to 50 mph. The Crane delivered 462,000 pounds of cargo five miles to shore in five hours as shown in Fig. 6.2.

When the Army needed its Cranes in Vietnam to offload an Australian aircraft carrier, which could not get to a loading dock, the Army asked Sikorsky what procedures and equipment were recommended. The answers came from the Skycrane experience in Bridgeport (Fig. 6.2). The Army benefited from this experience and was successful in offloading the carrier as explained in Chapter 5.

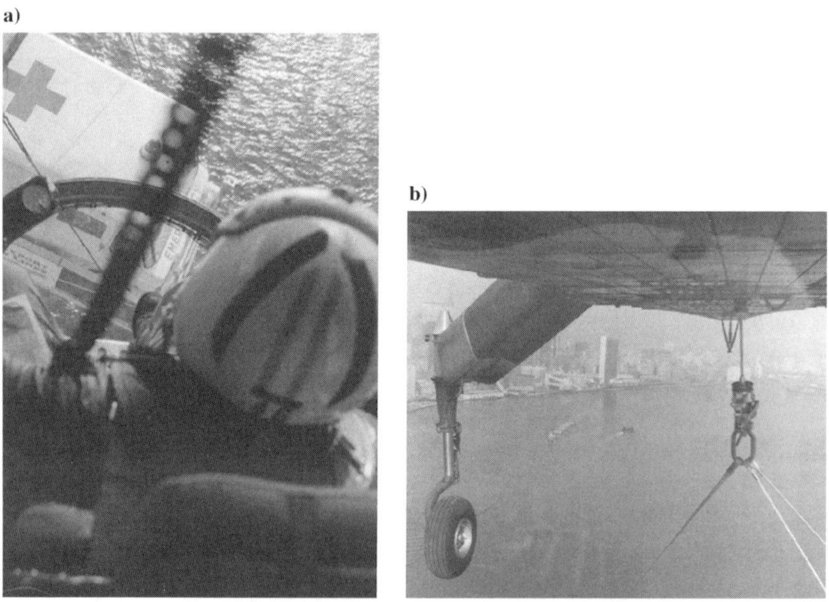

Fig. 6.1 Demonstration flight: a) carrying an emergency vehicle and b) main hook (foreground) and V-shaped tail skid (further back on the fuselage).

Fig. 6.2 Off-loading Isbrandtsen container ship near Bridgeport.

After this test, Isbrandtsen submitted a proposal to the Army to reduce the long waiting times (sometimes months) and added costs, which the Army experienced in offloading ships in Vietnam. It was a turnkey proposal using seven container ships and seven Skycranes, which would travel with the ships. In the proposal, the times for loading in America with Skycranes, for crossing the Pacific, and for unloading with Skycranes were based on Isbrandtsen's actual shipping and Skycrane experience.

The time saved by the Skycrane came from the actual Bridgeport test. The current shipping time for Isbrandtsen from the United States to Vietnam was taking 43½ days, including loading, shipping, waiting for a dock to unload, and unloading. Isbrandtsen was convinced that using the Skycranes would reduce the 43½ days to 20½ days, which would reduce the total time by 52.8% with a significant cost saving. The proposal was not accepted.

An analysis of the proposal reveals that the Army could have gotten much of that benefit if it used the CH-54s already in Vietnam to offload container-ships. The CH-54s could readily move loads from ships, waiting many days for a dock, to the correct Army base. This would eliminate the waiting time for the dock, waiting time for the offloading equipment, and waiting time for ground transportation to move the loads to an Army base. The local CH-54s could cut the total shipping time by 46% and also reduce costs. Such savings were confirmed when CH-54s offloaded an Australian carrier waiting for a dock in Vietnam as described in Chapter 5.

Fig. 6.3 Skycrane pulling loads from the bellies of Canadian resupply vessels.

SHIP TO SHORE FOR CANADIAN RESUPPLY

In another ship-to-shore demonstration, a Skycrane joined Canadian resupply vessels servicing remote northern villages, as shown in Fig. 6.3. This was not moving loads from the deck of the ship, but from a more difficult position, the hold within the belly of the ship. The Crane hovered over the ship, dropped its hook into the hold (where the crew attached it to the load), and flew the load directly to the storage site on land.

This technique was much faster than the usual method of the ship removing the loads from its hold and placing the loads on a lighter alongside the ship. The lighter was then moved to shore, where the loads were placed on the dock. Finally, the loads were loaded on trucks, which drove to the storage area for unloading there. All of this handling consumed time and cost. The Skycrane operation also enabled a major reduction in the typical spoilage and other losses of 25 to 35% experienced with the lighter operation to less than 2% with the Crane, a tremendous saving.

OIL PRODUCTION IN THE GULF OF MEXICO

Tug had many prospects in the Gulf of Mexico, where smaller helicopters were in use to support offshore drilling. The Skycrane flew there, arriving in December 1967. Tug invited representatives of the major oil companies and others who served the petroleum industry to inspect the Skycrane at a local airport. In two days, over 150 industry personnel visited the Skycrane, while Mal met with Chevron personnel, at their facility, to plan test flights.

Most of the customers thought the Skycrane was too expensive to operate, but Chevron did agree to let Tug use the Skycrane to fly a work-over sand rig to a drill platform 20 miles out in the Gulf as a demonstration. This was a scheduled maintenance operation for the platform.

The 52-ton rig was flown out in 11 trips, taking only one-and-one-half flight hours. Chevron said moving the rig by boat took 23 to 26 hours. During some of that time, production of oil was shut down. It appears the Skycrane would have reduced the production shutdown, but no data were available. The test did prove that the Skycrane could move loads rapidly and place them precisely, as shown in Fig. 6.4.

A few days later, Chevron called Tug with an emergency request. A platform was shut down and urgently needed an 8000-pound hydraulic pump replaced to restart production. It could not be moved the normal way by boat because of the weather.

Winds were gusting to 40 knots, and seas were four feet high. Chevron had already lost one day's production worth $10,800 and estimated it would be four more days before they could get the pump delivered by sea. In winds up to 50 knots, the Skycrane delivered the pump in 22 minutes, as shown in Fig. 6.5. The aircraft also placed the pump precisely where it was needed.

If the Skycrane had flown on the first day of bad weather saving five days of production, a rough estimate of the savings would be as follows: five days at $10,800/day equals $54,000 more oil produced in 1967 dollars or $378,000 in the estimated 2009 average dollars at $70 per barrel, minus the cost of 22 minutes of Skycrane flight time.

Oil Exploration in Columbia, South America

The team often had to develop new slings to manage the loads properly or other aids for new situations. As Mal says,

> When things didn't develop as planned, we would improvise in order to succeed. For example, when the ship delivering the Crane to Columbia, South America ran aground, the team put blades back on the Crane, started the engines, and began flying the drill rig sections from the stranded boat to the shore. Then the rig was flown piece-by-piece 50 miles into the impenetrable jungle to the drill site.

Mal concludes, "We performed a ship-to-shore mission first and then delivered the drill rig to the jungle site on schedule to meet the plan."

Fig. 6.4 Work-over sand rig engine assembly is delivered in the Gulf.

CONSTRUCTION OF NEW ORLEANS POWER LINE TOWER

New problems with different projects called for design changes to make the projects more feasible, as in this project that called for moving a power line tower in New Orleans. Mal explains,

> Often the adjustments to the operations plan were more extensive, as when we planned to disassemble a 312-foot-high electric power tower near New Orleans and reassemble it at a nearby location. During the first

Fig. 6.5 Skycrane delivers 8000-pound pump to a Chevron-producing platform in the Gulf of Mexico.

> reassembly test flight we found the tower sections continued to rotate due to the rotor downwash as the Skycrane attempted to put the tower section in place. I asked for an engineer from the plant to solve this major problem and we got Al Albert, an engineering genius at solving problems.
>
> He flew to New Orleans and quickly designed and built a cage for the cables which stopped the rotation of the tower.

One of the pilots, Lee Ramage, reports, "Al also designed and fabricated an angle guide that the aft pilot used to guide one tower section into another section. Al's angle guide system is used, with some modifications, on all transmission tower erections today."

Mal explains,

> Al's design allowed the Skycrane to precisely position the tower sections without help from ground personnel. Once a section was in place, ground personnel, already on the tower, would climb up to bolt the new section securely in place, while the Skycrane was picking up another section. This kind of action by the team was often needed to achieve success.

The Skycrane installed over 40 tons of steel tower sections reaching as high as 312 feet, all in 4 hours and 11 minutes, as shown in Fig. 6.6. If standard surface equipment had been used, it would have taken several weeks.

Fig. 6.6 Rebuilding 312-foot-high electric transmission tower near New Orleans; this is a 15,400-pound section.

Fig. 6.7 Building towers in a Virginia swamp.

Construction of Virginia Power Line Towers in a Swamp

Building a power line in a swamp can take huge amounts of money and time because it requires the construction of roads for the movement of heavy equipment from the beginning of the line to the end of the line. The Skycrane placed 48 five-ton towers, each 135 foot high for the Virginia Electric Company as shown in Fig. 6.7. It took the Skycrane only 20 minutes per tower, including the flight time to and from the staging area where the towers were assembled. No line men were needed on the tower to guide the tower sections in place because of the Crane's ability to place sections precisely.

Another benefit was less time from start to finish, yielding faster cash flow. The use of low-cost marsh or rough terrain reduced land cost and avoided populated areas. Access time to loads was greatly reduced by using the Skycrane, which reduced cost and time to complete. The result was a large savings in time and money. All of this was accomplished with only 18 Skycrane flight hours.

Construction of an Ocean Wall in Holland

Lee Ramage, one of the senior Skycrane pilots, relates "that in the fall and winter of 1971 and 1972, we flew the Skycrane in Holland to support the construction of a wall to hold back the ocean. The Skycrane picked up three concrete blocks in each load, weighing a total of 19,000 pounds. There were 4333 trips, which totaled 41,163 tons." Lee also says, "I would think the wall is still there."

Construction of Chesapeake Bay Bridge

Lee Ramage reports that the Skycrane was also used to pour over 14,000 yards of concrete for the new Chesapeake Bay bridge. He said, "The Skycrane was selected because the bridge was in very difficult place in the bay for a barge, with a crane on board, to pour concrete for the piers. The Skycrane completed this operation ahead of schedule and below budget."

Delivery of Heating Units to a 50-Acre New Factory Roof

In 1969, during the construction of a Chrysler Corporation plant, 263 heating units weighing up to 18,500 pounds each and measuring up to 12 ft × 20 ft × 12 ft had to be placed in many different positions on the 50-acre roof. The best conventional method was to reinforce the roof and erect a complicated and expensive system of cranes, dollies, and rails to move the heating units into place. This method placed an average of one unit per day and took a high number of labor hours. The Skycrane delivered the 263 units in just 6 days instead of 263 days, allowing the plant to open earlier while saving significant time and money.

COMMERCIAL VALUE PROVED

Mal's conclusion regarding the Skycrane demonstrations was that

> without the perseverance and determination of the Skycrane sales team and the support of Sikorsky engineering, the commercial Crane program would never have come to fruition. The major challenge was to prove to customers that the total cost of the Skycrane operation would save them time and money, compared to doing it conventionally. We proved that and the Skycrane sales followed.

COMMERCIAL APPLICATIONS

ALASKA OIL EXPLORATION BY ROWAN DRILLING COMPANY

In 1968, Rowan leased two Skycranes to support oil drilling on the North Slope of Alaska, an area of approximately 33,750 square miles (about 225 miles wide and 150 mile running north and south). Flying a shelter to the drill site is shown in Fig. 6.8. An article in the Winter 1970 United

Fig. 6.8 Delivery of construction shed to drill site.

Aircraft Corporation magazine *BEE-HIVE* describes the difficult operating conditions there.

> The two Skycranes arrived there in May 1968 after flying 4600 miles from the Sikorsky plant in Stratford, Connecticut. The helicopters were quickly put to heavy use.
>
> Rowan designed two drill rigs that broke down into sections to match the Skycrane lift capability. This greatly reduced the time to break down and rebuild these rigs, which were often moved to different exploration sites. The time saved resulted in a significant cost savings to Rowan and its customers, the oil companies, for which it was drilling.
>
> In the next three months, the two Skycranes flew 825 hours and carried millions of pounds of equipment and supplies. The material could not be delivered by ground transport because the tundra could not support trucks in the summer and the costs to build roads were excessive on the unstable tundra. One Skycrane flew 11½ hours in one day. A ten-wheel truck weighing 18,000 pounds was one load. The Skycranes were flying constantly from the Rowan base air strip to drill sites less than 20 miles away. Other drill sites were also planned to explore for oil on the vast North Slope.

With Skycranes flying over 11 hours a day when needed, the Sikorsky plant provided rapid supply of parts and assistance to keep the Skycranes flying. After many months of successful operation, Rowan reported that they had excessive wear and power loss on the Pratt and Whitney gas turbine engines. Rowan requested that a Sikorsky engineer fly to Alaska at once to inspect the engines and to find a solution to this serious problem.

The author arrived two days later, early one December morning with the temperature at minus 40 degrees and the wind blowing 50 mph. The ski clothes he wore were inadequate for this, and so he dashed from the small twin-engine Rowan aircraft to the nearest building on the Rowan gravel airstrip. One of the workers who greeted him said, as he looked out the window, "It's like seeing the surface of the moon." This certainly was an accurate description.

Examination of the engines confirmed the severe damage to the turbine blades, but gave no hint of the cause. The next step was to observe a Skycrane in its operating conditions, such as picking up material at the airstrip and delivering it to the drill site. From a smaller helicopter, the Skycrane was observed coming into a hover to deliver its load at a drill site. In the hover, the Skycrane was surrounded by a dense cloud of sand, which was used to stabilize the drill site. The engines were ingesting much of this material, which was clearly the cause of the engine damage. At the Sikorsky plant, it was thought the Skycrane was hovering over the more benign tundra or snow.

Back at the airstrip, the Rowan manager was told that the problem could be easily solved. The Army Cranes had this problem in Vietnam, and Sikorsky had developed an engine air particle separator (EAPS) that filtered out the sand in front of the engine intake. The EAPS caused only a small loss of engine power and a great increase in engine life. They could probably be shipped in a few weeks and could be installed by Rowan's own mechanics. The author was happy to leave that god-forsaken place after only four hours.

LOGGING

In 1971 Erickson Logging Company leased one Skycrane, which was flown out to Taylorsville, California, by a Sikorsky crew to demonstrate helicopter logging. The goal was to harvest high-value logs from deep-in woods that were very steep and isolated, but contained very high-quality wood.

This test would show that the Skycrane could open up these steep slopes for logging. The Skycrane would also eliminate the need for many new roads in the forests, which would avoid severe damage to the environment.

The trial was also to learn if the Skycrane could operate in a demanding logging production environment. The Skycrane proved it could be adapted to meet these goals, as shown in Fig. 6.9, and Erickson bought the first commercial Crane.

In the article "Flight of the Dragonfly," in the September 1998 issue of the magazine *Canadian Forest Industries*, Jean Sorensen saw the Crane as a huge dragonfly. He describes how the Erickson Air-Crane, Inc., logging system was applied to develop and expand logging into new areas:

> The men wait on a mountain ridge ... the ground tumbles down at a 70 to 80 degree slope.... It is beautiful scenery, but treacherous terrain. The logs are cut and trimmed ready for the Crane, which quickly comes to a hover and using the special equipment developed by Erickson, lifts the logs. A few minutes later, the Crane delivers the logs a short distance down the mountain for trucking to the mill.
>
> This is a way of getting steep slope wood that would not have been accessible or economical by any other means, the Dragonfly can pull wood from ground most men fear to tread. The speed with which the Erickson Crane system operates and the reliability of the Skycrane also contributed to the economical success, as the Dragonfly flew 10 and 12 hours a day when needed.

FIREFIGHTING

During the 1980s, Erickson began using the Skycrane for firefighting in the West Coast woods. Because the Skycrane was already logging in those woods, when called, the Crane could quickly remove its logging equipment at the base camp and fly to a nearby truck to attach the water pickup and delivery system before flying off to the fire. In the beginning, it was a large bucket,

a) **b)**

Fig. 6.9 Skycrane a) succeeds in first trail by Erickson and b) is purchased by Erickson for logging production.

which was dropped by the helicopter into a nearby lake or pond to get the water and then flown directly over the fire to drop the water as accurately as possible, as shown in Fig. 6.10.

To more fully utilize the Skycrane capacity, Erickson developed a sophisticated system, which included a 2650-gallon water tank attached to the Crane

Fig. 6.10 Using water bucket to attack the fire.

fuselage, as shown in Fig. 6.11. It can refill the tank in 45 seconds as the Crane cruises over a lake or hovered over a pond. Erickson calls this configuration the Helitanker and has sold a number of them. Today, they are operating on several continents. Operators such as Erickson, Evergreen, Sillers, and

Fig. 6.11 Erickson Helitanker with V-shaped water tank; the white hose on the ground is used to withdraw water from a pond in 45 seconds, while in a hover (courtesy of Erickson Group Ltd).

others have contracts with the U.S. Forestry Service (USFS) for use of their helicopters for firefighting on an as-needed basis.

In one of the largest fires in the history of Florida, which threatened one-third of the state a few years ago, helicopters throughout the United States were mobilized to fight it. Erickson sent five Helitankers to assist. Over a 25-day period, two of the Helitankers dropped over two million gallons of water and foam mix, averaging 19,000 gallons per hour.

As mentioned, Erickson's patented system allows quick refills of the tank in 45 seconds from shallow ponds and streams close to the fires. Because there were many rapidly spreading separate brush fires in Florida and many nearby ponds and streams, the Skycranes could refill and deliver another load of water in a few minutes to put out different brush fires. The Skycrane had a very quick cycle time, and so it was very productive with this capability.

AD CAMPAIGN

Sikorsky Aircraft supported the sales team effort with a series of ads. The first ad introduced "a revolutionary heavy-lift" helicopter "and the team that can make it pay for you." The ad states further:

> The continued growth of industrial/commercial transport requirements, especially in remote areas of the world, has created a need for new and specialized methods and equipment . . . it's the Sikorsky S-64E, 10-ton lift helicopter, supported by a widely experienced team of industrial transportation experts to help you and selected helicopter operators around the country.
>
> The military version of the SKYCRANE has been proven in over six years and 33,000 flight hours in heavy lift logistics support operations with the U.S. Army. . . . In a series of industrial development over the past two years in petroleum, shipping, power line, and construction industries, the SKYCRANE accomplished several heavy lift projects with significant saving in time, money, and manpower.

CONCLUSION

The initial applications of the Skycrane were mostly in logging, but most of them are now also used in firefighting and construction, as noted in this list of ten Skycranes sold commercially:

1968: Two leased to Rowan for oil drilling in the North Slope, Alaska
1971: One sold to Erickson for logging (originally leased)
1971: Three sold to Erickson for logging, construction, and firefighting
1975: Three sold to Evergreen for logging
1976: One sold to Tri-Eagle, a division of Louisiana-Pacific for logging

Ten does not add up to many commercial sales, but those ten were the only commercial Skycranes built due to concerns about the commercial demand for this expensive and unusual aircraft. The Skycrane sales team had two goals:

1) Define markets where the aircraft could operate profitably, such as logging, oil exploration, construction, and firefighting.

2) Find companies to become Skycrane operators to serve these new markets.

Today, several helicopter operating companies are using about 30 Skycranes. (Many of these helicopters were converted from retired Army CH-54s.) By this measure, the sales team did an outstanding job in helping to create new businesses, which is one of the most difficult of all marketing challenges.

This development of a commercial market recalls another quote from Igor Sikorsky: “In America … nothing can equal free work of free men. This is the foundation upon which the indisputable success of the United States has been built.” The sales team was free to target different industries, free to move about North and South America, free to define new uses, and free to design and apply new equipment and operating methods for the Skycrane to be successful.

Mal Burgess, former operations manager of the Sikorsky Skycrane sales team and former Vice President of Erickson Logging Company, says that “in both of my positions, the challenge was to create new Skycrane opportunities and achieve success.” It was difficult, but he and his team succeeded. The next chapter will explain how one company, Erickson Air-Crane, Inc., became a major Skycrane operator and innovator.

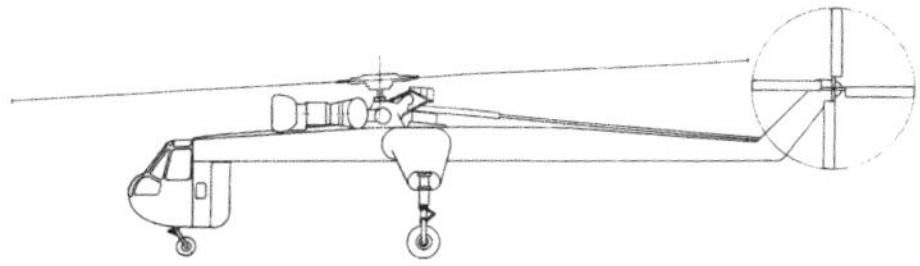

Chapter 7

Erickson Expands Markets and Buys All S-64 Rights

BY CLIVE WHITTENBURY*

Mr. Sikorsky's vision focused on a Skycrane design that would enable both military and commercial uses. The military made changes based on its experience and successfully applied them to the Skycrane. The commercial market needed an experienced innovator to provide the changes essential for commercial success. Jack Erickson was that innovator, and his story is told in this chapter.

Solving Problems in the Field the "Erickson Way"

Jack Erickson's father, Axel Erickson, was an innovative and successful timber logger. His specialty was accessing valuable timber through the use of an "aerial transport system" that had to be strung from the tops of trees to allow the logs to clear the ground on their way down from a ridge line to the road. The aerial system of heavy cables and pulleys was lifted to the top of a series of 100-feet-high trees by Axel, who climbed with the equipment. He had to cut all of the limbs off each tree as he climbed to the tree top.

From a very early age his son, Jack, operated the machinery that made this cable system work between mountainside and the roadside, where the logging trucks were located. Axel and Jack improvised solutions to difficult onsite logging problems just as they did for problems with their sawmill and truck fleet. That experience proved crucial to meeting future challenges of the Skycrane in remote locations.

Erickson and company met those challenges through what became known as the "Erickson Way" dealing with operating, mechanical, and other problems as they occurred. They drew on their experience to improvise as well as create innovative solutions on the spot. These abilities, with their years of experience with private aircraft operations, would aid in the success of the Skycrane. That is how Jack conceived the idea for helicopter logging with the

*Board member, Erickson Group Ltd. (Jack McKenna assisted in editing and Jack Erickson approved this chapter.)

Skycrane. The problems of operating and supporting helicopters in remote conditions awaited discovery.

First Erickson Helicopter Logging Trial

Wes Lematta owned a Sikorsky S-61 helicopter with enough payload (7000 lb) to try out the helicopter logging concept at Light's Creek, Taylorville, California, where Erickson had already been logging, using their aerial Skyline system. The sawmill was only eight miles from the logging site, which was on U.S. Forest Service (USFS) property and, therefore, required special permission to undertake the trial. The USFS was both cooperative and in the case of the Regional Forester, Dick Worthington, enthusiastic. He called it a "great idea" for the new USFS Environmental Initiative. Selective removal of valuable, old-growth timber was an objective of the USFS, as was reducing environmental damage by eliminating the need for new logging roads. The Skycrane would be essential to meeting both of these objectives.

The trial was demanding—an 8000-ft flight from the timber to landing the logs next to a creek-side road. Timber falling started in January 1971, and the flying test lasted three months. The payment for Wes Lematta's Sikorsky S-61 (Fig. 7.1) was "fuel costs plus a modest amount per measure of logs." Although this trial demonstrated basic feasibility, the S-61 payload was too small for moving the highly valuable logs. A payload of 20,000 lb looked much better for business than the 7000-lb payload of the S-61.

The results of the trial were discussed with the USFS who agreed that higher helicopter payload was crucial for economic feasibility. The USFS wanted to see this because it would open up new logging opportunities for the forest products industry. The USFS adopted four basic principles that would enable practical helicopter logging business opportunities:

1) At least 5-million board-feet need to be logged to justify setup costs.
2) There must be sufficient high-value timber to make it worthwhile.
3) There must be an open market for timber among the local mills.
4) Timber bidders must be qualified and have a helicopter already committed.

Fig. 7.1 Wes Lamatta's Columbia Helicopters Sikorsky S-61 used in first Erickson logging trial (courtesy of Erickson Group Ltd).

SKYCRANE TRIAL

The Skycrane had the needed payload, and five were sitting idle on the Sikorsky ramp at Stratford, Connecticut. Erickson leased one Skycrane for a logging demonstration at an Erickson logging site. The President of Sikorsky, Wes Kuhrt, answered an Erickson question about the possibility of problems with the Skycrane and the need for major expenses, such as component failure repair, with the statement: "If there is a failure, we fix it and our word is in our name." Jack Erickson responded by saying, "That's good enough for me." And that was the start of a simple and clear deal between Sikorsky and Erickson.

Erickson bid successfully on the first helicopter timber sale (the Drum Creek Sale) put up by the USFS and was ready for the Skycrane demonstration in the spring of 1971. The Skycrane was flown west by a Sikorsky team that later formed the core of Erickson Air Crane's flight operations, all with Sikorsky's encouragement and approval. On arrival with the Skycrane in May 1971, the Sikorsky flight team (Ramage, Evans, and Hoke) checked into a motel and went to look for food. On the main street they ran into Axel Erickson, whom they had met at Sikorsky. Axel took them to dinner and is said to have told the crew that "if the Skycrane will do what you say it will do, we will buy one."

The Drum Creek timber sale was operationally successful. It validated the concept for a helicopter larger than the S-61 and more clearly defined what could be expected from Skycrane helicopter logging. The Skycrane payload was about right, and all subsequent lumber sales were profitable through careful planning and Erickson's ability to deal with Skycrane problems quickly. Figure 7.2 shows several logs being lifted in a typical mountainous area.

"ERICKSON WAY" MEETS THE SKYCRANE AT DRUM CREEK

There was a problem at Drum Creek. The aft-facing pilot position did not work fast enough compared to the S-61 trial at Light's Creek. The Columbia helicopter pilot, Bob Brown, flew the S-61 and did pick-ups by looking out of the left pilot's side window. He was an able pilot, and so he could rapidly move the S-61 into position for picking up loads. The Skycrane had a rear-facing seat with limited flight controls, but this setup did not work fast enough for picking up logs, even with help on the ground. The pilot needed full control from the left seat.

The S-61 hooking procedure could work on the S-64, but logging from the front left pilot's seat was not possible in the Skycrane. The seat and collective (lift control) were too far from the left window. The answer was to move seat and control closer to the window. Erickson started work at shutdown and worked through the night to make the modifications. It was ready for flight the next morning at 7:00 am, including the approval by the

Fig. 7.2 S-64 moves logs from high-altitude slopes (courtesy of Erickson Group Ltd).

Federal Aviation Administration (FAA) of the changes. It worked better than the aft-facing pilot and as quickly as the S-61: Drum Creek was a success.

Skycrane's Second Logging Assignment

After the Drum Creek sale, Erickson won the Lobster Creek sale and committed the newly leased Skycrane to its first major production logging job. This was the beginning of profitable helicopter logging for Erickson. During this sale, Sikorsky's parent company, United Aircraft, took a strong interest in the Skycrane logging operation. The Sikorsky commercial sales manager, Tug Gustafson, who developed a visionary relationship with Erickson, together with Bill Gwinn and Art Smith (United Aircraft's Chairman and President, respectively) flew to Medford, Oregon, in the company's Jetstar to see the Lobster Creek operation. They drove into the woods, and the two cultures began communicating, effectively, onsite.

Erickson's Three Decades of Expansion

With the success of Skycrane logging, Erickson started what turned out to be a three-decade program consisting of the following:

1970s: Fixing Skycrane support problems and testing markets
1980s: Developing markets and Skycrane operations
1990s: Building an engineering and production base for expansion

The first decade began in the fall of 1971 with a visit to Sikorsky to discuss the purchase of all of the five Skycranes that were on the flight line. Agreement was reached on the Erickson boat, in Vancouver, British Columbia, in November of 1971, but then Erickson learned of the USFS finding that it could legally offer helicopter logging sales only with Skycrane competition.

Erickson found an independent buyer for one of the Skycranes, Evergreen, who would use it for logging. That meant Skycrane competition for logging right away from Evergreen. There was also potential competition from big forest product companies that could take orders at losing prices to get into the market against a much smaller Erickson Lumber Company. Competition was a clear risk for a small company, which had to buy expensive aircraft for use in an unproven market.

Maintenance, Repair, and Overhaul (MRO) Support for the Skycrane

The 1970s was a development period for the Erickson Skycranes. A new company, Erickson Air-Crane (EAC), was formed in Medford, Oregon, close to Erickson's main company. There were many aviation and timber specialists nearby to help develop, as needed, new operating and engineering solutions for the Skycrane. Although Sikorsky had committed to support the four Erickson Skycranes, their business procedures were not prepared for the demands of logging operations, with its short repair and turnaround times. High utilization rates of the Skycrane were essential for success. For example, refuelings without engine shutdown and flight cycles limited to 35 minutes between refuelings became standard.

Initially, Sikorsky could not meet Erickson's support needs in logging operations. But Sikorsky Engineering, which had produced an excellent basic helicopter design, worked well as Erickson pushed hard on the Skycrane. Some of the toughest conditions came from frequent engine power changes as the Skycrane picked up heavy loads, delivered them, and flew for another load every few minutes in a ten-hour day.

Dynamic flight loads on the Skycrane created large variations in stress during short-cycle lifting of full loads with elevation changes every few minutes. These conditions challenged major component lifetimes that, until then, were based on military operations, which normally had fewer cycles per day and milder maneuvers and flight loads. During one period, main gearboxes (MGBs) lasted for less than one hour of operation between failures, compared to the Sikorsky design time between overhaul (TBO) of 1300 hours.

It was first blamed on unforeseen operating conditions, but then it was found that the test stand for MGBs (at Sikorsky) was damaging newly overhauled MGBs as a result of overloading on the test stand. The solution, jointly defined by Sikorsky Engineering and Erickson, was to run-in overhauled MGBs for 30 minutes on a helicopter before first flight.

Logging operations created large variations in the stress on both airframe and components. Short cycles between the felled logs and the truck road, separated by hundreds of feet in altitude and subject to very tight placements, were the source of most of these large stresses and variations for the ten-hour day with a cycle time of three minutes. These flight conditions challenged the lifetimes of the major components. In a tribute to Sikorsky engineering, however, the flight envelopes were successfully extended, which was essential to commercial success. Less stressful operations were incurred in construction and logistics missions. Construction required extraordinary precision flying, for which the Skycrane had already become famous.

To replace major components, many smaller parts had to be removed and replaced frequently, causing delays. Erickson rearranged the smaller parts on the top of the helicopter fuselage, so that the major components could be more easily removed. These reengineering efforts helped to produce a very serviceable and highly reliable "Erickson Skycrane" by the end of the 1970s.

During the 1970s, the need for support from Sikorsky grew rapidly as operations expanded. This need had already imposed unplanned burdens on Sikorsky's repair and overhaul facilities, which had little reason to anticipate the rapidly increasing utilization of the commercial Skycrane. Fixing the

Fig. 7.3 This factory clean MGB is missing the hydraulic, electric, and other accessories that must be disconnected before the MGB is removed from the aircraft (courtesy of Erickson Group Ltd).

lifetime problems of the major and minor components took meticulous work over several years.

The MRO (maintenance, repair, and overhaul) issues revolved around direct support for the Skycrane, where failures of the major component and systems (MGB, engine components and accessories, flight systems, blades, and main rotor head) could quickly reduce job and business performance.

Failure rates for the MGB are an excellent example of those difficult times. The rated time between overhaul was 1300 hours, but failures were occurring in about 100 hours. These failures were "normal" as opposed to the rash of few-minute failures already referred to. Many failures required removal and replacement of the MGB in the middle of the forest (often at night)—a process requiring two days. The total weight of the MGB is over 1500 pounds. It was a tremendous challenge to disconnect a nonfunctioning MGB, remove it, replace it with another, and reconnect the new one to driveshafts, electrical and hydraulic outputs, etc., in the woods at night. This was a serious business challenge, and so Erickson developed solutions. A new MGB is shown in Fig. 7.3.

The engine shown in Fig. 7.4 also had problems. When an engine developed a seal leak, instead of sending it back to Pratt and Whitney, a log loader was employed to remove the engine from the Skycrane and suspend the engine in the air while the power turbine was replaced. The Skycrane was back to logging the next day. The engine was another very large component with many connections that often had to be repaired at night in the woods.

By rearranging the parts on the top of the fuselage, where the MGB, radiator, engines, auxiliary power unit, and hydraulic stand and accessories were installed, Erickson reduced MGB replacement time from two days to six hours for a reduction of 75%. These and other actions improved the actual

Fig. 7.4 One of the Pratt and Whitney JFTD12 gas turbine engines on the Skycrane (courtesy of Erickson Group Ltd).

time-to-failure of the MGB from 140 hours to the TBO of 1,400 hours. That is an improvement by a factor of ten. The MGB, fuel control, and parts durability problems were greatly improved by the end of the 1970s.

REBUILDING THE FIRST SKYCRANE

In 1976, Erickson started rebuilding a Skycrane, damaged in an accident, by fabricating new airframe sections and using parts available from spare airframe assemblies. Because Erickson had learned a good deal from its MRO experiences, this rebuilding also presented an opportunity to modify the Skycrane to solve a number of the MRO problems. As an example, the transmission oil cooler was removed from the transmission and remounted on the fuselage deck, allowing easier access to the transmission for removal.

George Howard's wisdom, during preliminary design, in eliminating fairings and enclosures on the top deck of the fuselage (see Chapter 3) proved invaluable. Relocating parts on the engine deck reduced repair times, saving precious minutes and hours during intense logging operations. During this process, Erickson obtained FAA approval of over 30 engineering modifications on the Skycrane.

A flight accident with one Skycrane, late in 1974, and a later crash from an engine failure with another Skycrane reduced the Erickson fleet to two and put additional pressure on Sikorsky support to assist in their repair and return to flying status. Sikorsky had built three more Skycranes at the end of Army CH-54 production. EAC offered to purchase all of them, but was refused by the new Sikorsky president. It appears this refusal was because new Sikorsky management did not want to increase the MRO workload from Erickson into the Sikorsky product support department. Two of those Skycranes went to Evergreen to give it three, and one went to TriEagle, a subsidiary of Louisiana-Pacific, a large forest products company.

Sikorsky would not support the repair of the damaged Skycrane, and so Erickson used its deep experience to proceed on its own. Having engineered and certified many improvements in maintenance and operations, Erickson was in a good position to remanufacture its own improved version of the Skycrane. The rebuild of the first crashed Skycrane was completed in May 1977. This Skycrane started at Erickson with 1913 hours flying time and, thirty years later, had exceeded 38,000 flying hours—attesting to a quality repair.

Although the 1970s were a trying time for everyone, the result was a triumph for Erickson, for Mr. Sikorsky's vision, and for the Sikorsky engineers who designed and supported the Skycrane. Mal Burgess, who was the operations manager for Tug Gustafson on the Sikorsky sales team in the 1960s, became vice president of operations at Erickson in the 1970s. He relates, "Though relations with Sikorsky management were difficult at times,

Jack Erickson continued to work well with Sikorsky key individuals. I think that is why there were three Erickson Skycranes, with one of these names painted on each of their noses, IGOR, TUG, and MAL."

INITIAL DIVERSIFICATION OF ERICKSON MARKETS

Construction projects changed the flying demands on the Skycrane from high-production frequent lifting (logging) to precision flying of very valuable loads where the aft-facing pilot was critical. Projects included placement of loads on tall buildings, solving access problems on factory roofs, and building power transmission lines in difficult terrain, while avoiding the cost of making roads. Erickson was fortunate in seeking and finding innovative customers from caisson cement pouring for the Chesapeake Bay Bridge to building ski lifts (Eastern and Midwestern United States). Erickson was selected in 1975 to place the antennas on the CN tower in Toronto, Canada, which, at the time, was the world's tallest self-supported building. The operation required a new design of the load guide by Erickson to enable the lift and accurate placement of 36 sections without assistance of workmen in the tower itself. Use of the Skycrane took only 3-1/2 weeks instead of the 6 months needed for conventional construction. Figure 7.5 shows the Skycrane placing one of the 36 sections of the antennas on the tower.

Erickson was selected to remove and return the refurbished Freedom statute from the top of the U.S. Capitol building (Fig. 7.6). President Clinton

a) b)

Fig. 7.5 Views of the CN tower in Toronto: a) placing antennas on top and b) close-up view (courtesy of Erickson Group Ltd).

Fig. 7.6 Placing the Freedom statue on the U.S. Capitol building (courtesy of Erickson Group Ltd).

and key members of Congress were there in person, while thousands watched from the Mall and many others throughout the United States via CSPAN. This event was the culmination of quality engineering, trust in Erickson, and a vindication of Igor Sikorsky's vision.

Erickson responded to another major, high-pressure project to restore Juneau, Alaska's power supply, which was interrupted by a fall storm in 1975 that blew down a transmission line in the mountains above Juneau. A new right-of-way had to be logged to clear the way for a new line. The project became known through a movie "Race Against Winter," as Erickson won that race for Juneau (Fig. 7.7).

In 1975, Erickson organized a subsidiary to provide transmission line construction for the Bonneville Power Administration (BPA). It also performed building construction work across the eastern United States. This type of work had longer flight cycles, less dynamically demanding flight loads, and the need for precision flying by the aft-facing pilot, which allowed this pilot to precisely deliver the load.

Large transmission lines were put in place at the rate of three to four miles per day, with individual precision lift cycles of 15 minutes, three sections to a tower and four towers per mile. All this was done without human hands on

Fig. 7.7 Building the new Juneau electric power line (courtesy of Erickson Group Ltd).

the steel. Erickson had completed 9000 miles of 500 kV transmission lines within just a few years.

The Skycrane had, by the end of the 1970s, successfully demonstrated its capabilities in several different markets, and it had achieved a reliability record much enhanced by the Erickson team. One Skycrane had also been successfully rebuilt by Erickson.

Engines, which originally went to TBOs of 400 hours, had now reached 3500 hours, and MGBs had improved from 640 hours to TBOs to 2100 hours. Another key component, the rotor blades, had moved from a lifetime of 3000 hours to a spectacular lifetime of 20,000 hours through clever and innovative residual stress engineering undertaken at the Erickson MRO facility. This was an improvement of almost seven times.

GLOBAL MARKET GROWTH IN THE 1980S

Erickson needed to rebuild its second damaged Skycrane to expand globally. Its badly damaged airframe required critical data support from Sikorsky. The new management at Sikorsky was supportive, but only for that data vitally important to the rebuild. Careful selection and restriction of data made this agreement work effectively.

Erickson built a new tail boom and pylon from scratch, while a salvaged fuselage center section and cockpit were restored and mated using key interface data from Sikorsky. It took over one year to completely restore the helicopter, with a successful test flight in May 1980. It went to work in remote areas in northern British Columbia and later in Indonesia and Malaysia. This new Erickson version of the Skycrane proved equal to the heavy workload in remote mountain areas.

Summary of the 1980s

Erickson had, by the end of the 1980s, established a full MRO capability that enabled the reengineering of the Skycrane for reliability, the complete rebuilding of damaged Skycranes, and the full logistics support of operating Skycranes in remote global deployments. This had been achieved using only four Skycranes in several different markets. This management challenge required the development of profitable flying and ground-support operations to ensure success for Erickson.

Global Growth in the 1900s as Erickson Buys All Skycrane Rights

Successful operations in the 1980s opened a new market—forest firefighting. This market led to significant growth in the Skycrane fleet assisted during the 1990s when Sikorsky agreed to sell the Skycrane Type Certificate to Erickson in 1992. This transferred all drawings, tooling, and spares with all manufacturing and support rights worldwide.

Sikorsky kept the name Skycrane, and Erickson adopted the name Air-Crane for the S-64. This agreement gave Erickson the design rights to both the S-64E (10-ton payload) and S-64F (12.5-ton payload) versions. Rather than manufacturing new airframes, Erickson planned to build an Air-Crane fleet through converting military CH-54A and CH-54B, both of which were being sold as Army surplus.

Many of these airframes and spares went to different bidders. Erickson took a different route. Erickson was restoring a number of historically significant aircraft that could be traded through museums for CH-54s. Restoration preserved national treasures for aviation history while receiving in return parts of the CH-54 surplus. The first treasure was a flyable Sopwith Camel, a World War I fighter used by the United States, for a surplus CH-54 at the Army Museum at Fort Rucker. The original parts for that Sopwith came from Tommy Sopwith's garage in the United Kingdom, so the aircraft was a valuable bit of history.

Price of a Remanufactured CH-54 Skycrane Reaches $28 Million in 2009

As the surplus Army Skycranes became scarce, the price for a remanufactured, in good condition and updated Erickson Aircrane, with all engineering modifications was about $30 million by "street estimates." The sale of one in mid-2009 was leaked at $28 million. The last three of the surplused CH-54s may have moved into remanufacture at this writing.

The first Army CH-54s, several decades before, cost the Army about $2,800,000 each, as reported in Chapter 3, which included armor for pilots and engines, self-sealing fuel tanks, military electronics, and other military equipment. The simpler commercial version cost may have been closer to $2,600,000. This is less than one-tenth the 2009 price of $28 million, even after many years of inflation, an increasing interest in the commercial market, and a machine that has now operated for long periods with high reliability.

The commercial value increased as the capabilities and reliability improved. Forty-five-year-old aircraft that are not continuously upgraded do not demand ten times their original cost. However, Skycrane value increases because of the extensive improvements that have been made continuously for two decades. The viability of continuing development work clearly shows the inherent genius and value of Igor Sikorsky's concept and is a tribute to the original engineering layout that made these improvements possible. The development of the Skycrane for extended and heavy production use, yielding high commercial productivity, has justified increasing value and price. Early in Canadian logging, for example, 300 hours were flown in 30 days using double crews to average 10 hours per day. One commercial Skycrane fuselage is reported at 25,000 flight hours.

Use of surplused CH-54 airframes and parts expanded the Erickson fleet by another 10 Erickson Air-Cranes. The Erickson fleet climbed to 14 Air-Cranes by 1995 and climbed again in the 2000s to a total of 18. Additional Air-Cranes have been sold to South Korea and Italy (Fig. 7.8) for firefighting and other uses.

FIREFIGHTING EXPANSION IN THE 1990S

By 1994, Erickson was logging in Canada and Malaysia with a major portion of its growing fleet and rapidly advancing the Air-Crane's capability to fight fires. Buckets were used originally to drop water on fires. Erickson developed a specially shaped 2650-gallon tank for accurate spreading or aiming of water, foam mix, or retardant in eight different delivery streams as shown in Fig. 7.9.

The Erickson tank uses two methods for refilling. A patented ram-scoop hydrofoil refills the tank in 40 seconds as the Air-Crane cruises above a small lake or pond. The Air-Crane can hover and refill its tank in 45 seconds over water only 18 inches deep using a snorkel with a high-speed impeller. The tank is computer-controlled and is patented, and it can be added readily to an Air-Crane. In this configuration the Air-Crane is called the Helitanker (Fig. 7.10).

Fig. 7.8 Two Air-Cranes sold to the Italian government; the first is configured for general purpose and the second for firefighting (courtesy of Erickson Group Ltd).

a) **b)**

Fig. 7.9 Air-Crane's displaying firefighting capabilities: a) tank drops water on fire and b) water cannon fires water precisely (courtesy of Erickson Group Ltd).

a)

b)

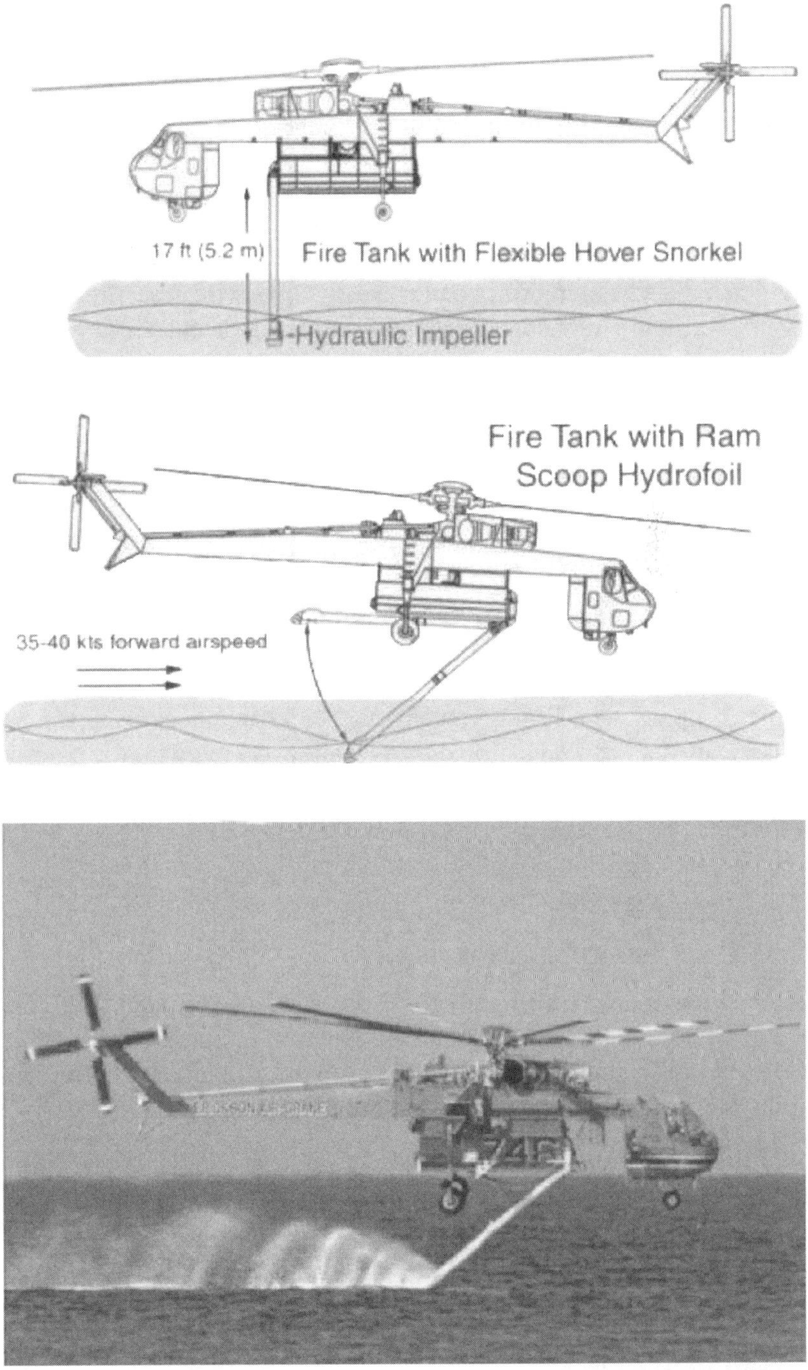

Fig. 7.10 Helitanker: a) refilling the water tank and b) using the scoop (courtesy of Erickson Group Ltd).

Helitankers have fought fires worldwide, including Australia, South Korea, Malaysia, Brunei, Italy, Turkey, Greece, France, and Canada. In Brunei, in one month, two Helitankers flew over eight hours per day dropping over 12 million gallons of water on fires. Bruneian fire authorities estimate the Helitankers assisted in controlling 95% of the inferno.

The fire caused a deadly haze, which posed a health threat to nearby citizens. Because of the smoke, the fixed-wing firefighting aircraft could not safely fly low, but the Air-Cranes could make low approaches to the hot spots underneath the advancing smoke to attack the fires with great precision. These attributes, carefully developed over years, have made the Air-Crane so useful worldwide.

Innovation from Two Cultures

There was both clash and cooperation between the two very different organizations, Sikorsky and Erickson. Sikorsky was a large defense contractor, with over 10,000 employees, learning how to support one of its few commercial products: a radical helicopter designed for new markets. Sikorsky's management and commercial product support policies and supervision changed several times in this period. Erickson was a small, successful, fast-moving company with limited resources. Erickson made the Skycrane yield profits in new markets on a global basis while working with the colossus that was Sikorsky Aircraft. The opportunity for misunderstandings between the companies was expected, but key people on both sides endured to ensure success.

Sale of the Erickson Air-Crane Company

Holding the manufacturing and logistics rights to the Air-Crane, Erickson expanded its global capabilities to the Americas, Europe, and Asia. There have always been discussions about the need to build more Air-Cranes for use in Asia and Europe. This would imply major growth for Erickson Air-Crane itself. Jack Erickson sold his company to an investor group in 1997 to enable that globally oriented new opportunity. Thirteen years later, the exciting story of Erickson Air Crane continues to unfold.

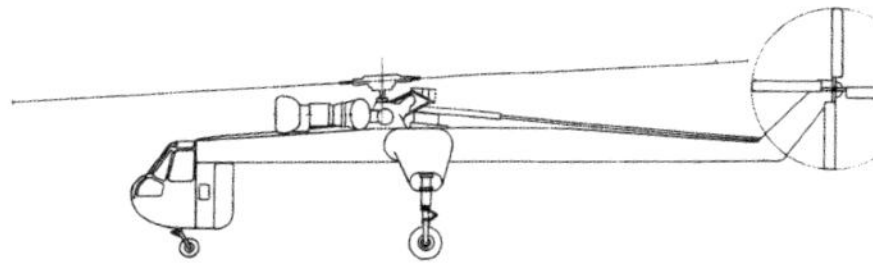

Chapter 8

Future Cranes

When will a more powerful Crane be developed? Several design studies for a more capable Crane exist in the Sikorsky Archives, with a design developed 45 years ago.

1964 Study

The earliest Sikorsky study defined a growth version, which appears slightly larger than the first S-64, but with much higher payload. The installed engine power was 11,000 horsepower from two engines. The rotor diameter was increased from 72 to 79 feet, resulting in a gross weight of 65,700 pounds versus 38,000 pounds for the first S-64 (see Fig. 8.1); 42,000 pounds for the CH-54A; and 47,000 pounds for the CH-54B. At delivery of the first Army Crane in 1963, gross weight was 42,000 pounds and about five years later

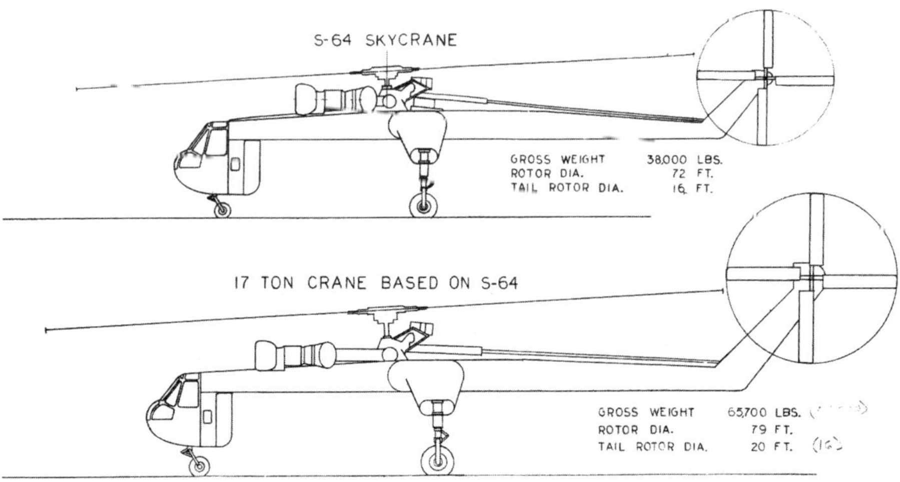

Fig. 8.1 Top) 1961 S-64A 38,000-pound gross weight; bottom) 1964 design for a larger Crane.

47,000 pounds for the later version of the Army Crane. The 1964 study yielded a payload of 34,000 pounds or 17 tons versus 25,000 pounds or 12.5 tons for the CH-54B Skycrane.

1970 COMPETITION

The Sikorsky response to an Army-defined requirement in 1970 for a heavy-lift helicopter with a payload of 22.5 tons was powered by three General Electric TF34/58 engines delivering 7000 horsepower each at maximum power for ten minutes at sea level at 95 degrees Fahrenheit. It had a four-bladed 124-feet-diameter main rotor and a four-bladed 26-feet-diameter tail rotor.

Design gross weight was 118,055 pounds. The artist rendering shown in Fig. 8.2 is correct for this design, except that the artist shows a different engine option, which is four engines shown mounted on the sides of the fuselage. The three engines of the design were mounted on the top of the fuselage, two in front and one to the rear of the MGB. Otherwise, the sketch is an accurate representation of this proposal to the Army.

This competition was won by BoeingVertol, which also proposed an 118,000-gross weight design, but using tandem main rotors. A development contract was awarded in 1971 for the Boeing Vertol XCH-62A, but the contract was cancelled in 1972, before the first helicopter was completed.

Fig. 8.2 Artist sketch of four-engine version of 1970 Crane design.

CH-53K CRANE

A lower-cost and lower-risk approach to develop a more capable Crane would be to develop it from the CH-53K now being designed for the Marine Corps. The CH-53K is a much improved, higher-performance version of the CH-53E, which first flew in 1974 and is performing valuable service in the Navy and Marine Corps. The CH-53K is being developed at this writing at Sikorsky Aircraft on a $2 billion contract.

One hundred fifty-six aircraft are planned for the Navy and Marine Corps, with an in-service target date of 2015. Maximum gross weight is 74,000 pounds with internal loads, and maximum gross with external loads is 84,700 pounds. It employs a 79-feet-diameter main rotor with seven blades. Total horsepower planned from three engines would be 22,500 shp. A picture of the predecessor CH-53E is shown in Fig. 8.3.

The commercial Crane version of the CH-53K would eliminate the military equipment such as aerial refueling, armor, large external fuel tanks, blade-folding mechanism, tail-cone folding structure, military electronics, cargo rail locking system, and other special Marine equipment. The large cabin would be replaced with a simple stick fuselage, and the fairings and work platforms for engines, MGB, etc., would be eliminated to save weight,

Fig. 8.3 Three-engine CH-53E, first flown in 1974, the predecessor of the CH-53K.

increase payload, and enhance maintainability. Because most commercial and Army Crane missions have been five to twenty miles, the increased drag caused by exposed components would be offset by increased payload, reduced maintenance, and reduced aircraft downtime. These changes for the Crane would reduce the estimated CH-53K flyaway cost of $56.6 million, published in *Rotor & Wing* magazine, by 15% to $48.11 million for the commercial CH-53K Crane, according to one calculation.

Additions for the Crane version would be an aft-facing pilot cockpit, wide landing gears to clear large-sized loads, and a hoist to lift 20 or more tons. The 53K Crane would benefit from almost 50 years of technology improvements because the first Skycrane was designed in 1961. For example, the 53K Crane would retain the CH-53K high-efficiency rotor blades with anhedral tips, low-maintenance elastomeric rotor head, joint interoperable glass panel instruments, fly-by-wire flight controls, composite structure fuselage, and other improvements.

This configuration is aimed at the commercial market, where about 30 Skycranes are operating today on a worldwide basis. One study forecasts demand for 100 Cranes for future markets. The Skycranes have established these commercial markets, but Skycranes will require more maintenance as they become much older. The first S-64 flew in 1962 and will be 52 years old in 2014. A 53K Crane could be developed most economically after the CH-53K is developed. This could be at a very low development cost compared to developing a new Crane from scratch.

It will also benefit from the military support and improvement system to be set up for the CH-53K. This system normally operates for twenty years or more for a new military model helicopter and usually provides benefits such as design improvements and lower spare parts costs. For all of these reasons, the 53K Crane would be a low-cost way to develop a more productive new Skycrane.

One approach to estimating the payload of the 53K Crane is to use the actual weight empty-to-gross-weight ratio of 41.9% of the commercial S-64F, which had a 25,000-pound payload or 12.5 tons. The CH-53K gross weight of 84,700 pounds times 41.9% yields a weight empty of 35,489 pounds. If the weight of crew of 660 pounds, trapped fluid weight of 500 pounds and one hour of fuel of 8000 pounds (for series of short-range Crane missions) were added, this would yield a payload of 40,051 pounds or 20 tons. Three General Electric GE38-1B engines would deliver 7500 shp each to a seven-bladed main rotor and four-bladed tail rotor.

Lee Jacobson, retired Sikorsky director of aircraft design and development, has forecast that the Navy/Marine CH-53K could reach 100,000 pounds gross weight, perhaps as soon as five years after first flight, which is forecast for November 2011 at the time of this writing. His growth estimate is based on his large database, which is the growth in gross weight of five earlier Sikorsky helicopter types, over a period of 40 years.

Applying Lee Jacobson's Gross Weight Growth Theory to the S-64, we find the initial gross weight was 38,000 pounds as shown in Fig. 8.1. This first Army Crane was delivered with a 42,000-pound capability and later with 47,000 pounds for a total increase of almost 24%. If the 53K Crane, or perhaps better named Skycrane II, gross weight of 84,700 increased by 24%, it would yield about 105,000 pounds for a mature Skycrane II. A 24% increase in payload would increase payload from 20 to about 25 tons, which would meet many of the Army requirements.

ARMY 53K CRANE

If the 53K Crane were developed and applied successfully in the commercial market for some years, it might even be considered for Army procurement. The 25-ton payload 53K Crane could lift almost all of the Army heavy equipment, except for the largest weapons, such as the main battle tank. The original Skycrane was demonstrated in the commercial marketplace for several years before the Army bought it, and so it could happen again because such a Crane can move more types of cargo faster than any other helicopter and it can also perform many critical missions that no other helicopter can perform.

HELPFUL RELATIONSHIPS OF THE S-64/CH-54 AND THE S-65/CH-53

It has been said that history often repeats itself. That seems to apply to the S-65/CH-53K providing the basis of an improved Crane. This would be in reverse to the S-64/CH-54 preceding the CH-53A in its initial development. The S-64A Skycrane first flew in May 1962 with a 72-feet-diameter, six-bladed main rotor and in a few years with gross weight of 42,000 pounds. The CH-53A/S-65 first flew for the Marine Corps in October 1964, two years and five months later than the S-64A, with a 72-feet-diameter, six-bladed main rotor and a gross weight of 42,000 pounds.

The S-64A Skycrane flew to an Army base to demonstrate its capabilities one month after first flight and then to the Marine Corps. It was actively engaged in this manner for over two years before the CH-53A flew. The CH-53A design most likely benefited from the flight testing of S-64A with the Army and Marine Corps. Certainly, a 53K Crane design, being developed after the CH-53K, would benefit from the flight testing of the CH-53K, hence the possible helpful relationship.

COMPARISON OF ACTUAL AND PROPOSED CRANES

Table 8.1 summarizes the main rotor diameters, gross weights, and payloads of the Cranes that have been built, the ones proposed, and CH-53K.

TABLE 8.1 COMPARISON OF ROTORS, WEIGHTS, AND PAYLOADS OF CRANES

Year	Model	Rotor, diam	Gross, weight	Payload, tons
1964	CH-54A delivered	72	42,000	10
1968	CH-54B delivered	72	47,000	12.5
1964	Growth S-64 concept	79	65,700	17
1970	Design for Army quote	79	118,055	22.5
2009	CH-53K Crane concept	79	84,700	20
2019	CH-53K Crane growth	79	105,000	25

The year 2019 is obtained by adding the four years of growth for the CH-54A to CH-54B to the planned operational date of the CH-53K of 2015.

This comparison reveals the tremendous impact of technology improvements on helicopter performance. The 84,700-pound gross weight of the CH-53K Crane concept has about 89% of the payload (20 tons/22.5 tons) of the 1970 concept of 118,055 pounds, and it weighs 28% less (84,700 pounds/118,055 pounds).

The performance of the 79-feet-diameter main rotor increases from the 1964 Crane concept lifting 65,700 pounds to the 2009 CH-53 Crane concept lifting 84,700 pounds for a 29% improvement. Much of this benefit is caused by more powerful engines and improved blade design.

CONCLUSION

Many believe an improved Crane is needed. Lieutenant Colonel (retired) Heffner, a former commander of the 273D Assault Support Helicopter Company (Heavy) in Vietnam provided his views in Chapter 5. Lieutenant Colonel (retired) Brock, a former commander of the 478th Aviation Company of CH-54s in Vietnam, was quoted in an undated Army publication as follows: "The future of the Heavy Lift Helicopter is limited only by the funds and our imagination. It promises exciting possibilities that we did not dream of a few short years ago."

At this time, there is no program to develop a more capable Crane to replace the Skycrane. Perhaps this book will assist in bringing that about. What may be required is a person or company with the vision and passion to lead the effort to build on Igor Sikorsky's vision.

Index

Supporting Materials

Many of the topics introduced in this book are discussed in more detail in other AIAA publications. For a complete listing of titles in the AIAA Library of Flight Series, as well as other AIAA publications, please visit www.aiaa.org.

AIAA is committed to devoting resources to the education of both practicing and future aerospace professionals. In 1996, the AIAA Foundation was founded. Its programs enhance scientific literacy and advance the arts and sciences of aerospace. For more information, please visit www.aiaafoundation.org.